Macmillan Encyclopedia of the Environment

GENERAL EDITOR

STEPHEN R. KELLERT, Ph.D.
School of Forestry and Environmental Studies
Yale University

ASSOCIATE EDITORS

MATTHEW BLACK, M.E.S.
Resident Naturalist
Connecticut Audubon Coastal Center

RICHARD HALEY, M.E.S.
Former Assistant Director
New Canaan Nature Center
New Canaan, Connecticut

ADVISERS

DORIS CELARIER
Public Affairs, U.S. Forest Service
U.S. Department of Agriculture

LINDA FROSCHAUER
Former Middle Level Director
National Science Teachers Association

JOYCE GROSS
Public Affairs
U.S. Department of Commerce/NOAA

VALERIE HARMS
Consultant, National Audubon Society

DAVID McCALLEY
Assistant Director
Center for Energy and Environmental Education
University of Northern Iowa

JAMES PRATT
Environmental Sciences and Resources
Portland State University

MARION SADER
Librarian
The Hackley School
New York

JOHN TANACREDI
Supervisory Ecologist
National Park Service
U.S. Department of the Interior

ANNE TWEED
Environmental Science Teacher
Eaglecrest High School
Colorado

Macmillan Encyclopedia of the
ENVIRONMENT

VOLUME 2

General Editor
Stephen R. Kellert

Associate Editors
Matthew Black

Richard Haley

Macmillan Library Reference USA
New York

Developed, Designed, and Produced by Book Builders Incorporated

Macmillan Library Reference
1633 Broadway, New York, NY 10019-6785

Library of Congress Catalog Card Number: 96-29045

Printed in the United States of America

Library of Congress Cataloging-in-Publication Data

Macmillan encyclopedia of the environment.
      p.   cm.
    "General editor, Stephen R. Kellert"—P. iii.
    Includes bibliographical references and index.
    Summary: Provides basic information about such topics as minerals, energy resources, pollution, soils and erosion, wildlife and extinction, agriculture, the ocean, wilderness, hazardous wastes, population, environmental laws, ecology, and evolution.
    ISBN 0-02-897381-X (set)
    1. Environmental sciences—Dictionaries, Juvenile.
    [1. Environmental protection—Dictionaries. 2. Ecology—Dictionaries.]
    I. Kellert, Stephen R.   96-29045
GE10.M33 1997   CIP
333.7—dc20   AC

Photo credits are gratefully acknowledged in a special listing in Volume 6, page 102.

D

Dams

▶**A**rtificial structures built across streams to partly block the water flow. The building of dams is a widespread activity that has occurred for thousands of years. A dam partly blocks a stream to create a buildup of water in a RESERVOIR behind the dam. Most dams have many uses. For example, the water in the reservoir formed by a dam may be used for recreation, for drinking, for irrigating farmland, for generating HYDROELECTRIC POWER, or for all these purposes. Dams are sometimes built to regulate the flow of a stream and to prevent flooding below the dam. More than 15% of the surface RUNOFF on Earth ends up in a reservoir behind a dam. Most dams achieve their goal of controlling river flow. In many parts of the world, millions of people depend on dams for their survival and their employment.

While dams have many practical uses, they can also create problems. Often, they produce a whole series of problems that their designers may or may not have predicted. For example, dams slow water flow and often flood valleys. In this process, dams often destroy wildlife HABITAT. Dams may also ruin sports such as whitewater rafting by eliminating fast-moving water along a river. Such problems have often resulted in bitter fights between conservationists and pro-dam people.

When a dam is constructed, the river's flow is controlled, altering the natural conditions in the river. For instance, the water temperature in a reservoir is generally warmer than that of a free-flowing river. At the same time, more water is lost by evaporation from the surface of the reservoir than from the smaller surface of a river.

◆ The Hoover Dam, on the Arizona-Nevada border, controls the floods of the Colorado River. It supplies electricity and irrigation water to a large area of the southwestern region of the United States.

The construction of a dam changes the characteristics of habitats both above and below the dam. Living things may be affected in many ways by these changes. For example, existing land habitats are shrunken in size or destroyed when they are flooded by the waters of a reservoir. Animals living in such habitats are forced to leave to survive. Similarly, dams prevent fish from migrating freely up the river. In addition, the reservoir creates still-water habitats. Such habitats may be favorable to many flies and snails that transmit diseases.

Dams have many geological

◆ The Aswan High Dam controls the floodwaters of the Nile River in Egypt. Lake Nasser was built to hold surplus water as the Nile rises during the rainy season.

effects. SOIL and plant nutrients are often trapped behind dams. After a dam is constructed, fields and habitats below the dam receive fewer nutrients and less SEDIMENT than the river once carried. If the coast downstream is eroding away, as it is at the mouth of the Mississippi River and many other North American rivers, the coast erodes faster after the dam is built.

In older and poorly maintained dams, the buildup of silt behind a dam may cause the dam to burst. If a dam bursts, the area below the dam may become flooded. Often, many people may have settled in areas below dams. If severe flooding occurs as a result of dam failure, much property and many lives may be lost.

In the area above the dam, the WATER TABLE rises. This may cause soil to become waterlogged and salinize. Some scientists believe changes in water pressure above and below a dam may even cause earthquakes.

A dam alters the local CLIMATE. The reservoir causes humidity levels to increase and temperatures to fall in areas above the dam. As a result, PRECIPITATION increases above a dam and decreases downstream.

Most major waterways that can be dammed in the United States have dams now. Because of the controversy and legislation, such as the WILD AND SCENIC RIVERS ACT, few additional large dams are likely to be constructed. Other countries will likely continue to build some dams to provide reservoirs of drinking water and supply small hydroelectric plants with energy. [*See also* BONNEVILLE POWER ADMINISTRATION; BUREAU OF RECLAMATION; and TENNESSEE VALLEY AUTHORITY.]

Darwin, Charles Robert (1809–1882)

British scientist and naturalist who helped develop the theory of EVOLUTION. Darwin wrote in his book, *On the Origin of Species by Means of Natural Selection,* "There is a frequently recurring struggle for existence.... Any being if it vary however slightly in any manner profitable to itself will have a better chance of surviving and thus be naturally selected. This preservation of favorable individual differences and variation, and the destruction of those that are injurious, I have called Natural Selection or the Survival of the Fittest." Thus, as the fittest survived and passed on their favorable individual differences to their offspring, the animal or PLANT gradually changed or evolved.

Charles Darwin was born in Shrewsbury, Shropshire, England, on February 12, 1809, into a well-to-do family. He attended Cambridge University, where he prepared to become a clergyman. Upon graduation from college he was asked to sign on as naturalist on the sailing ship HMS *Beagle,* which was about to embark on a surveying expedition of the Southern Hemisphere. Darwin had always been interested in animals, and he was eager to learn about those SPECIES that lived in the southern part of the world. The five-year voyage took him to South America, Tahiti, New Zealand, Australia, Tasmania, and many different lands.

STUDYING EARTH'S SURFACE

Darwin's observations during his travels convinced him that Earth was a changing, evolving place, a view not shared by many at the time. The *Beagle* was sailing off the coast of Chile when a damaging earthquake occurred. When the ship arrived in port, the crew saw the terrible destruction of lives and buildings. Darwin noticed that the land had risen by 3 feet (1 meter) or more. He believed this showed that Earth was undergoing changes. He thought that the center of Earth was made up of molten rock, which could break through Earth's surface and change it.

Later, by studying CORAL REEFS, Darwin found evidence that Earth's crust sometimes sank. He knew that coral could not live at ocean depths below 120 feet (37 meters). When he found dead coral below that level, Darwin theorized that the land level was sinking. His observations about Earth's changing surface helped prepare him for the important studies about life that he was to make.

CHANGING ANIMALS

At Punta Alta on the southeastern coast of South America, Darwin made an interesting discovery. He found the fossils of giant animals. They were larger versions of familiar animals—sloths, armadillos, hippopotamuses, ELEPHANTS, and llamas. These findings made him wonder if living things changed through the ages.

It was in the GALÁPAGOS ISLANDS off the coast of Ecuador where Darwin made his most important discovery. There Darwin found species of PLANTS, REPTILES, BIRDS, and INSECTS that were slightly different on each island, and sometimes different on the same island. The most striking was several species of birds called finches. The different finch species in the Galápagos are similar, but each has a different beak size and shape. Some have heavy beaks for cracking seeds; others have thin beaks for pulling insects from holes.

◆ The warbler finch (left) has a small bill and eats only insects. The large ground finch (right) has a large bill and eats hard seeds. It rarely eats insects.

It seemed to Darwin that the finches on the Galápagos had developed from a single finch species that had probably been blown to the island by a storm. From these observations, Darwin developed a theory that animals developed variations to suit their ENVIRONMENT.

HIS WRITINGS

Darwin arrived back in England in 1836 and began to write about his five years of observation and study. In 1839, he married his cousin Emma Wedgwood and together they had ten children.

Darwin published many works in his time. His writings include *Journal of Researches, The Structure and Distribution of Coral Reefs, The Variation of Animals and Plants Under Domestication, The Descent of Man,* as well as *The Different Forms of Flowers on Plants of the Same Species,* and many other works about plants.

Darwin's most famous book, *On the Origin of Species,* explains his theory of NATURAL SELECTION, the idea that Earth is constantly evolving as well as his theory of evolution—that we are all descended from common ancestors. This theory "provided a foundation for the entire structure of modern biology," said fellow scientist Julian Huxley. *The Origin of Species* sold out in several days and has been referred to as the "book that shook the world" because of Darwin's controversial theories.

◆ The Galápagos tortoise has changed through the ages.

Darwin worked on his research and writings up until two days before his death. He died on April 19, 1882. He was buried in England's historic Westminster Abbey, resting place of British heroes. [*See also* CONTINENTAL DRIFT; PLATE TECTONICS; and VOLCANISM.]

DDT

❱D ichlorodiphenyl trichloroethane (DDT), a chemical INSECTICIDE that contains CARBON, hydrogen, and chlorine. It is a poison that acts by paralyzing INSECTS. DDT was first synthesized in 1874 by the German chemist Othmar Zeidler. In 1939, it was shown to be an effective insecticide by a Swiss chemist, Paul Muller, who received the Nobel Prize for his work nine years later.

Soon after the United States entered World War II, the DEPARTMENT OF AGRICULTURE assigned a group of scientists the task of finding a way to protect military and civilian personnel from insect-transmitted diseases. The team found DDT to be highly effective against disease-carrying insects. The colorless, odorless substance was easy to make and could be prepared as a powder for dusting or as an oil or water solution for spraying. Although it was deadly to a large variety of insects, including mosquitoes, lice, fleas, flies, Japanese beetles, bedbugs, and termites, it was only slightly harmful to MAMMALS. Because it did not break down

when exposed to sunlight or as a result of chemical reactions in SOIL or the bodies of animals, it remained in the ENVIRONMENT to kill any insect that encountered it.

The U.S. Army's first well-publicized use of DDT was in February 1944, when it ended an epidemic of **typhus fever** in Naples. Afterward, it was widely used and is credited with eliminating **malaria** and **yellow fever** in certain parts of the world. Edward F. Knipling, a scientist with the Department OF Agriculture, estimates that DDT and similar compounds have saved 50 million lives.

By the late 1950s, however, many experts had become concerned about the widespread use of these poisons. Although DDT had a devastating effect on insect populations at first, some individuals survived and passed on their natural resistance to their offspring. By the late 1960s, more than 200 kinds of insects were known to have become

◆ Sometimes planes are used to spray pesticides.

resistant to DDT and similar compounds.

Scientists also discovered that DDT sprayed on PLANTS and into water moved through the FOOD CHAIN. It traveled from primary CONSUMERS—the animals that ate the plants—to secondary and tertiary consumers. Secondary consumers are animals that eat primary consumers, and tertiary consumers are animals that eat secondary consumers.

As DDT moved up the food chain, it accumulated in the bodies of animals. The accumulation had harmful effects. FISH and small BIRDS died. In larger birds, DDT disrupted the function of chemicals called *steroids*, which play an important role in producing eggs. After they began to lay thin-shelled eggs that did not hatch, birds such as the BALD EAGLE, peregrine falcon, brown pelican, and osprey suffered sharp population declines.

◆ Pesticides used to get rid of caterpillars may harm the birds that eat the caterpillars.

Quaternary Consumers
DDT in fish-eating birds (ospreys)
25 ppm

Tertiary Consumers
DDT in large fish
(needle fish) 2 ppm

Secondary Consumers
DDT in small fish
(minnows) 0.5 ppm

Primary Consumers
DDT in zooplankton
0.4 ppm

DDT in water
0.000003 ppm
or 3 ppt

◆ The concentration of DDT increases as it goes up the food chain. The amount of DDT in water or in the tissue of an organism is measured in parts per million (ppm).

In the late 1950s, the Department of Agriculture became concerned by evidence that DDT was getting into food. To keep DDT out of milk, the Department of Agriculture called for a ban on using the insecticide on cows or their feed. In addition, DDT could not be used on lettuce, broccoli, and other edible plants that cannot be peeled before being eaten.

In 1962, a biologist named Rachel Louise CARSON sounded the alarm in her book *Silent Spring*. This best-selling work awakened Americans to the dangers of DDT. In 1969, Robert Finch, Secretary of Health, Education, and Welfare, recommended that most use of DDT be phased out in the United States within two years. Today, its use is banned or severely restricted in most developed countries. As a result, the populations of bald eagles and other sea birds are increasing.

Insecticides have been developed that do not linger and accumulate in the environment. However, the future of insect control may lie with a system called INTEGRATED PEST MANAGEMENT (IPM). IPM calls for the use of BIOLOGICAL CONTROLS, such as BACTERIA and other natural enemies, that attack insects. Substances that prevent the insects from maturing or reproducing are also being explored. Only when these approaches fail are measured amounts of insecticides used. [*See also* AGRICULTURAL POLLUTION; AGROECOLOGY; BIOACCUMULATION; COEVOLUTION; FEDERAL INSECTICIDE, FUNGICIDE, AND RODENTICIDE ACT (FIFRA); HEALTH AND DISEASE; NATURAL SELECTION; PEST CONTROL; PESTICIDE; POLLUTION; and WATER POLLUTION.]

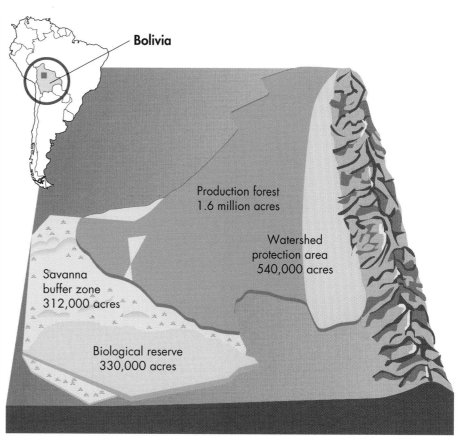

◆ A conservation organization in the United States made a debt for nature swap to help protect a tropical forest area in Bolivia from destructive development.

Debt for Nature Swap

▮A financial agreement between a debtor nation and a conservation organization in which a debt is canceled in exchange for an environmental protection program that saves or restores threatened parts of the ENVIRONMENT and encourages a sustainable environment. The idea of debt for nature swaps was first suggested in 1984 by biologist

Thomas Lovejoy, assistant secretary for external affairs at the Smithsonian Institution in Washington, DC.

Developing countries borrow large amounts of money from banks and the industrialized nations of the world. Collectively, these debts are rapidly approaching $1 trillion. With interest, the debts increase yearly. In an attempt to meet their financial obligations, many developing countries are exporting lumber and abusing land resources at a rate that is devastating their NATURAL RESOURCES. Despite these attempts, world economists generally agree that it is unlikely that many developing countries can repay their huge debts.

Huge financial debts combined with the destruction of the environment make developing countries ideal candidates for debt for nature swaps. Private environmental organizations such as Conservation International and the World Wildlife Fund purchase the debt of debtor countries in the form of bonds from banks at a greatly discounted rate. They then use the bonds to develop environmental conservation programs in the debtor country. In exchange for such programs, the debt of the country is cancelled.

The first debt for nature swap took place in 1987 between Conservation International and the government of Bolivia. Using funds from a private foundation, Conservation International purchased $650,000 of Bolivia's $5.7 billion debt from a bank in the United States for a discounted price of $100,000. In return for Conservation International's cancellation of the debt, Bolivia agreed to create a fund to manage and protect 3.7 million acres (1.5 million hectares) surrounding the Beni Biosphere Reserve, a tropical forest region in the Amazon basin.

In 1989, the United States government made a grant of $1 million to the World Wildlife Fund. The money was used to purchase more than $2 million of Madagascar's debt. In return, Madagascar agreed to begin reforestation programs and increase and train new park rangers. Other countries involved in debt for nature swaps include Costa Rica, Brazil, Peru, Mexico, Jamaica, and Tanzania.

The idea of debt exchange, especially in developing nations, where funds for repaying debts may not be available, is not new. Earlier debt exchange projects include raising funds for youth, agriculture, education, health, and other important social programs. Debt exchange programs offer advantages to the debtor country, conservation organizations, and the world. By cancelling the debt of the debtor country, that country now has the potential to develop a sustainable society leading to a global economy. Environmental conservation programs reduce AIR POLLUTION, WATER POLLUTION, and DEFORESTATION; conserve natural resources; and restore BIODIVERSITY.

Although the concept and prac-tice of the debt for nature swap is considered a step in the right direction, it raises some important issues. In addition to incentives and commitment on the part of the debtor nation, questions of monitoring and enforcement have been raised. One suggested solution is the creation of agencies to monitor and assist in carrying out such programs. These same agencies may be empowered to reduce the debtor nation's financial obligation over time as the project moves ahead.

Deciduous Forest

▶ A temperate area of wooded land dominated by broad-leafed trees that drop their leaves each autumn. Such trees include maple, oak, beech, birch, and others having broad, flat leaves. People who live near deciduous forests know the beauty of the red, amber, and gold colors taken on by the leaves of these trees during the autumn months. This change of color results from a loss of chlorophyll, their green pigment, just before they die and fall to the ground.

The shedding of leaves by deciduous trees is an ADAPTATION that helps these trees survive the winter months. The scarcity in winter of sunlight and liquid water needed in photosynthesis prevents deciduous trees from absorbing enough energy to maintain their life processes as well as their leaves. Dropping leaves helps deciduous trees conserve water during this time. The fallen leaves also help the trees survive by insulating their roots. The leaves that remain on the ground may also prevent it from freezing solid.

During winter, deciduous trees are in a state of dormancy, or rest, that is similar to the hibernation periods of some animals. They drop their leaves, stop growing, and use the little energy available to them just to survive. In spring, when light and water become more available, trees use the additional energy to produce the leaves, flowers, seeds, and fruits needed to maintain the life of individual trees as well as the survival of tree SPECIES.

Deciduous forests are located throughout North America, Europe, and Asia in regions between 30° and 60° north latitude. CLIMATE conditions in these regions are generally described as **temperate**, but they may vary greatly over such large regions. For example, the availability of light, water, and other resources may allow for a growing season that lasts between four and six months. Summer temperatures may be mild or may soar to greater than 95° F (35° C). Winter temperatures are often well below freezing. Deciduous forests are generally moist, receiving between 30 and 100 inches (75 and 250 centimeters) of PRECIPITATION annually. Depending upon season and temperatures, precipitation may fall as rain, sleet, hail or snow.

The temperatures and moisture levels of deciduous forests support a large variety of DECOMPOSERS. These organisms—mostly BACTERIA and FUNGI, together with other protists—help the mounds of fallen leaves and other dead organic matter decay rapidly. As a result of this process of decay, the SOIL of deciduous forests tends to be rich in nutrients.

Like tropical RAIN FORESTS, deciduous forests grow in distinct layers. The upper layer, or canopy, is formed from the many leaves and branches of trees such as maple, oak, and birch that dominate many deciduous forests. Beneath the tall trees grow a variety of shrubs and bushes. These plants form the **understory**. Sunlight that peeks through the leaves and reaches the forest floor supports the growth of a variety of FERNS, herbs, and mosses.

In addition to their diverse PLANT life, temperate deciduous forests also support a variety of other organisms. Although the exact species of organisms vary greatly from one deciduous forest to another, depending upon where in the world the forest is located, most deciduous forests provide HABITAT for a large number of animal species, representative of each major animal group. Common to many deciduous forests of the United States are large MAMMALS such as black bear and deer, as well as

smaller mammals such as mountain lions, foxes, squirrels, and chipmunks. Snakes and tortoises are among the REPTILES that make their homes in these forests. AMPHIBIAN

◆ Spring

◆ Summer

◆ Fall

◆ Winter

◆ The different seasons in a temperate deciduous forest are shown above.

populations may include frogs, sala-manders, and toads. A variety of BIRDS, INSECTS, ticks, spiders, worms, snails, and other animal groups are also present. [*See also* BIODIVERSITY; BIOME; CONIFEROUS FOREST; DECOMPOSITION; DEFORESTATION; and OLD-GROWTH FOREST.]

Decomposer

◗ An organism that obtains energy by feeding on animal wastes and the remains of organisms. As it feeds, a decomposer releases nutrients that can be recycled through the ECOSYSTEM back into the ENVIRONMENT.

FUNGI and many types of BACTERIA are decomposers. Decomposers feed on organic matter such as animal wastes and the remains of dead PLANTS and animals. Such wastes are known as DETRITUS. Thus, decomposers are sometimes called *detritivores.*

Because decomposers obtain their food from other organisms, they are a type of CONSUMER. However, decomposers differ from other consumers because they digest their food before taking it in.

To obtain nutrients, fungi and bacteria release chemicals into the organic matter that serves as their food. These chemicals carry out digestion by breaking down complex molecules into simpler molecules. Once digestion has taken place, the decomposer takes in the nutrients it needs from the matter. Nutrients not taken in by the decomposer are left in the environment,

◆ Earthworms eat soil and digest organic matter using useful material and getting rid of undigestable matter.

where they can be used by other organisms. In this way, decomposers help to recycle matter through the ecosystem.

Decomposers occupy the final position in all FOOD CHAINS and FOOD WEBS. In this NICHE, decomposers play a vital role in the recycling of food and chemicals throughout the

ecosystem. For example, the nutrients released from organic matter through DECOMPOSITION are often released into SOIL. Plants take in the nutrients from soil and use them for growth. As these nutrients are taken in by consumers that feed upon plants, the nutrients are again cycled through the ecosystem by way of

◆ Mushrooms and other fungi decompose dead plant material, such as this branch.

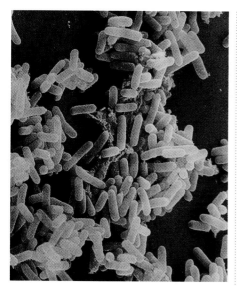

◆ Bacteria, seen here under a microscope, are important decomposers.

food chains and food webs. [*See also* BIODEGRADABLE; BIOGEOCHEMICAL CYCLE; CHEMICAL CYCLES; COMPOSTING; HUMUS; and TOPSOIL.]

Decomposition

▶ The breakdown of organic waste material—plant and animal matter— by organisms. Organisms called DECOMPOSERS feed on the sugars, starches, and proteins present in animal wastes and in the remains of organisms.

Decomposers include many types of BACTERIA and almost all FUNGI. Decomposers often feed near the ground, where the remains of organisms tend to accumulate. Here, decomposers carry out two important ecological roles. The first is to release important nutrients back into

the ENVIRONMENT. This task is accomplished through the feeding action of decomposers.

As they feed, decomposers release digestive enzymes into the organic matter that serves as their food source. These enzymes break down complex molecules into simpler substances. The decomposer then takes in the nutrients it needs for its survival. The remaining nutrients remain in the air or SOIL, where they can be taken in and used by other organisms. In this way, decomposers help recycle chemical substances in the environment.

◆ Humus is an end product of decomposition.

The second role of decomposers is to build new soil. The soil material produced by decomposers is called HUMUS. Humus is partially decayed plant and animal matter that makes up the uppermost layer of soil. It generally has a dark color and spongy texture. Because humus

contains many of the nutrients released by the decomposition process, it is a good soil for the growth of plants. The spongy texture of humus also aids the growth of plants by allowing this soil material to hold water and air. Together, humus and rock particles make up the layer of soil known as TOPSOIL. [*See also* BIOGEOCHEMICAL CYCLE; CHEMICAL CYCLES; COMPOSTING; and WEATHERING.]

Deep Ecology

▶ A social and political movement that calls for a rethinking of the way people relate to and view the ENVIRONMENT. The term deep ecology was introduced in 1972 by Arne Naess, the Norwegian philosopher. However, Aldo LEOPOLD is known as the father of deep ecology. In 1949, a collection of Leopold's essays entitled, *A Sand County Almanac,* was published. In one of the essays, "A Land Ethic," he pointed to the need for establishing a system of values or moral attitudes about how land is used.

Leopold's writings have greatly influenced the deep ecology movement. The basic philosophy of deep ecology views nature as having a true, national value. Thus, nature should be viewed independently of human needs and the value humans place on NATURAL RESOURCES. Another defining feature of deep ecology is that BIODIVERSITY is, in and of itself, an expression of the beauty and multitude of life forms.

Deep ecologists criticize and challenge the view of environmental groups who claim to work toward protecting the environment but do very little to bring about change. Naess and others characterize such practices as a compromise. They view the idea that resource management and pollution are concerns only as they relate to human health and welfare as "shallow ecology."

that used these areas as HABITAT died. Thus far, the forest and mangrove regions destroyed by the defoliants have not returned to their original condition. Many ecologists believe the area will never recover because other plants have colonized the deforested areas.

The use of defoliants in Vietnam had a severe impact on the ENVIRONMENT. However, many people believe the defoliants, especially Agent Orange, which contains DIOXIN, caused severe illnesses in

people exposed to the chemicals. Newspaper accounts from Saigon claimed that people living in areas near where the defoliants were used suffered severe illnesses shortly after chemicals were sprayed. Among the problems were high rates of miscarriages and birth defects in children. Soldiers who were in the region at the time the chemicals were used developed such symptoms as headaches, diarrhea, nausea, and severe skin rashes. Many of these soldiers later developed serious, long-term

Defoliant

▌A type of HERBICIDE specifically designed to remove the leaves of plants. Herbicides are chemical substances that are used to rid an area of unwanted vegetation. Defoliants are applied to plants by spraying or dusting.

During the 1960s, the United States became involved in the civil war in Vietnam. To prevent troops from being harmed by enemy soldiers hiding in the dense foliage of the FORESTS and MANGROVE swamps of Vietnam, the United States used helicopters, planes, and boats to spray the region with defoliants. The three defoliants that were used were Agent Orange, Agent White, and Agent Blue.

The use of these defoliants in Vietnam completely destroyed about 5% of the DECIDUOUS FORESTS on which they were sprayed. Almost half of the country's mangrove swamps were destroyed. In addition, many FISH and other animals

◆ During the Vietnam War, the chemical Agent Orange was sprayed on dense vegetation in Vietnam. Today, little remains of the original rain forest.

disorders that they believe are caused by the defoliant. Although the United States banned the use of Agent Orange partly because of these complaints, the controversy about whether or not it and other defoliants caused the illnesses continues. [*See also* CANCER; CARCINOGEN; DEFORESTATION; HEALTH AND DISEASE; and PESTICIDES.]

Deforestation

▶The destruction of FORESTS for the purposes of AGRICULTURE, timber, FUEL WOOD, and cattle ranching. Defor-

estation is most severe in the tropical RAIN FORESTS of Africa, Asia, and Central and South America, where about 60,000 to 80,000 square miles (16 million to 20 million hectares) of forest are cleared annually.

Over half the world's tropical rain forests have already been destroyed by deforestation. About 2% of the remaining forests—an area about the size of Florida—are cleared each year. At this rate, scientists predict that no rain forests will remain by the middle of the twenty-first century.

Deforestation is also occurring in other countries, including the United States. Although the clearing of forests serves many purposes for humans, scientists are concerned about the short-term and long-term effects of deforestation on the ENVIRONMENT.

CAUSES OF DEFORESTATION

Deforestation is a consequence of human POPULATION GROWTH. As the world's population grows, increasing amounts of land and resources are needed to support it. Most of this growth is now occurring in developing nations, such as Brazil, Peru, Zaire, and Indonesia. These regions also contain the largest tropical rain forests. In fact, tropical countries with the highest population growth rates usually have the highest deforestation rates. Today, people clear forests for a variety of reasons.

Agriculture

Agriculture accounts for most of the world's deforestation, totalling about 80,000 square miles (200,000 square kilometers) annually. When forests are cleared, space is opened up for the planting of crops. In many countries, large-scale **plantations** are rapidly replacing tropical forests. In Central America, for instance, where coffee and bananas are the main cash crops, farms now stand where forests once existed.

In tropical rain forests, most land for agriculture is cleared by the slash-and-burn technique. Slash-and-burn agriculture is an old and productive practice for clearing land in order to grow crops in nutrient-poor SOIL. Despite the abundance of lush vegetation, tropical rain forests have surprisingly poor soil, because most of the nutrients available in tropical rain forests are in the organisms that live there, not in the soil.

To replenish the soils with nutrients, a farmer clears a small plot of

◆ In Costa Rica, many forest areas have been cleared extensively.

◆ The large-scale deforestation of the Amazon rain forest is a major concern of enviornmentalists.

Cattle Ranching

Tens of thousands of square miles of tropical rain forests are also cleared annually to make room for cattle **ranching**. In South and Central America alone, some 20,000 square miles (52,000 square kilometers) of forest, an area the size of New Jersey, are cleared each year to create

land by cutting the vegetation and then burning it. The ashes that remain act as an organic fertilizer, enriching the soil so crops can grow. Crops are grown on the plot of land until the nutrients are used up. The farmer then moves to another patch of forest.

Logging for Timber

Each year, about 50,000 square miles (130,000 square kilometers) of tropical rain forest are cut for timber, paper pulp, and other wood products. Trees are harvested for a variety of reasons. They are used to make paper and furniture, as well as the lumber and plywood needed for the construction of homes and other buildings. Harvested trees also provide firewood for many people. In fact, about 1.5 billion people in developing countries depend on firewood as their major source of FUEL. However, much of the timber from tropical rain forests is sold to other countries. Timber **exports** provide many developing countries with essential income.

There are many different methods for harvesting trees. CLEAR-CUTTING, the process of removing all of the trees from a land area, is the most popular method of tree removal because it is the least expensive. Clear-cutting is the method of choice in many developing nations. However, some tropical countries have taken action to curb the loss of their forests. Some nations, such as Panama, have made it illegal to clear-cut forests. These nations recommend selective cutting, in which only middle-aged and mature trees are cut down. Often, seedlings are then planted to replace those trees that are removed.

pastures for GRAZING. Once cleared for cattle ranching, a patch of land is only useful for a short period of time. After a few years, soils are eroded and compacted by the movements of vehicles and cattle. Eventually, the land is abandoned and the rancher clears a new patch of forest.

CONSEQUENCES OF DEFORESTATION

When a tropical rain forest is destroyed, much more than trees is lost. HABITATS are destroyed and countless animal and plant SPECIES are pushed to the brink of EXTINCTION, and INDIGENOUS PEOPLES are displaced and their sources of livelihood are lost. In addition, soils become infertile and suffer EROSION. When deforestation is extensive, local and global CLIMATES can be altered.

Loss of Biodiversity

Although they cover less than 10% of the world's landmass, tropical rain forests support almost 65% of Earth's organisms. Thus, scientists are concerned that deforestation is contributing to a great loss in BIODIVERSITY. Deforestation can lead to species extinction directly through habitat destruction, as well as indirectly. Tropical rain forests are extremely complex ECOSYSTEMS in which species depend on one another for a variety of reasons. Scientists fear that the extinction of one species can affect many other species in a chain-reaction fashion. For instance, the disappearance of a pollinating animal, such as a fruit bat, can cause the extinction of the plants they pollinate. In turn, the extinction of these plants can affect animals that feed on them.

Scientists are also concerned that as the rain forests are destroyed, species are becoming extinct much faster than they can be identified and studied. Many of these species, particularly plants, could supply humans with valuable food products or chemicals that could be used to make new drugs and medicines.

Destruction of the Soils and Climate Change

Extensive deforestation can also lead to changes in soil quality and local climate. In tropical rain forests, most nutrients are stored in living and decaying organisms, rather than in the soil. When trees are cleared, soil erosion increases and the already nutrient-poor soils become even more infertile. The hot, tropical sun bakes exposed soil, forming a hard, brittle surface, which can be more easily eroded by rainfall.

Deforestation can also lead to changes in the local climate. Tropical rain forests control regional climates by heating and cooling the air, soil, and water, maintaining **humidity,** and offering protection from the wind. When these forests are cleared, an important source of water is also lost. Eventually, infertile soils and drought can lead to DESERTIFICATION of an area.

Global Climate Change

Scientists also fear that continued deforestation of the world's tropical rain forests may contribute to the GREENHOUSE EFFECT and GLOBAL WARMING. The loss of trees can lead to the greenhouse effect in two ways. First, when trees are burned to clear land, they release large amounts of CARBON DIOXIDE (CO_2), an important GREENHOUSE GAS, into the ATMOSPHERE. In fact, it is estimated that more than 1 billion tons (1 billion metric tons) of CO_2 are released annually from

◆ Tropical rain forests are being destroyed at an alarming rate as the human population increases.

◆ Many of the rain forests in Indonesia are being cleared for agricultural purposes.

SLOWING DEFORESTATION

Currently, scientists are working vigorously to slow and reverse current rates of deforestation. These efforts include establishing tropical forest preserves to protect forested land, improving management of unprotected forests, improving agricultural techniques, and curbing the demand for tropical forest products. [*See also* AGROECOLOGY; BIOME; CARBON CYCLE; CLIMATE CHANGE; ECOSYSTEM; ENDANGERED SPECIES; HABITAT LOSS; and RESTORATION BIOLOGY.]

burning forests. Second, the loss of trees and other plants means that less CO_2 is being absorbed from the atmosphere through PHOTOSYNTHESIS. Overall, deforestation leads to increases in CO_2 levels in the atmosphere. With increased levels of CO_2, scientists believe more heat is trapped near Earth's surface, causing global warming.

Demography

◗ The study of the distribution and makeup of populations. Population experts, called *demographers*, analyze information about living beings that exist in a certain area, using data from the national population census or from national surveys. In the United States, the national population census is taken every ten years.

HUMAN POPULATION STUDIES

In studying the human population of a community, ECOSYSTEM, city, state, region, or country, demographers collect and study data about all aspects of people's lives. For example, they may study age, sex, race, income, employment, and level of education. Statistics about lifestyles, hobbies, and personal property are also examined.

◆ This forest has been damaged by acid rain.

Demographers may study whether people are married or single; have children; smoke; drink; or own a car, gun, CD player, TV, boat, motorcycle, recreation vehicle, or computer. They may determine how many people hunt, fish, go camping, ski, bird-watch, sew, swim, golf, play tennis, or take part in other outdoor activities. They study the books and magazines people read and whether they belong to or support environmental groups such as Greenpeace or the World Wildlife Fund. The analysis of such data, called *demographics*, helps manufacturers know where to advertise products or build businesses.

The United Population Institute analyzes worldwide statistics. The information is used and studied by countries and the United Nations. The Cairo Population Conference studies population trends and makes recommendations relative to population problems.

STUDIES OF NATURAL POPULATIONS

Demographers also track human population changes—births, deaths, MIGRATION, and variations in population density. Ecological demographers track these same population changes for the WILDLIFE that make up ecosystems. Analysis of such data helps scientists know if a SPECIES is in danger of EXTINCTION and where to place wildlife preserves. Analyzing both human and wildlife demographics also helps scientists predict potential problems caused by human activities. [*See also* AGE STRUCTURE; OVERPOPULATION; and POPULATION GROWTH.]

Deoxyribonucleic Acid

See DNA

Department of Agriculture (USDA)

❚ The executive department of the U.S. government that works to ensure ample supplies of farm products, expand overseas trade of farm products, and guarantee reasonable profits for farmers while assuring consumers fair prices for these farm products. The U.S. Department of Agriculture (USDA) also certifies the suitable condition of food supplies by inspecting meat and poultry in slaughtering and processing plants; by grading meat, poultry, and dairy products to verify their quality; by administering a nationwide system of grain inspection; and by estab-

WILLIAMSTOWN HUNTER SURVEY						
Age	Sex	Marital Status	Do You Hunt?		Own a Gun?	
26	male	married	yes	no	yes	no
22	female	single	yes	no	yes	no
46	male	married	yes	no	yes	no
37	female	married	yes	no	yes	no
20	male	single	yes	no	yes	no
40	male	married	yes	no	yes	no
31	male	single	yes	no	yes	no
62	male	divorced	yes	no	yes	no
52	female	married	yes	no	yes	no
26	female	divorced	yes	no	yes	no
35	male	married	yes	no	yes	no
29	female	single	yes	no	yes	no

◆ By studying data about people, demographers can better predict their needs. Similarly, by studying data collected in the field by researchers, scientists can better protect wildlife in an ecosystem.

lishing standards of quality for any grain exported from the United States. The department is headed by the secretary of agriculture, a member of the president's cabinet, who is appointed by the president, with Senate approval.

AGRICULTURAL AID

The USDA finances research at its own labs and at agricultural experiment stations of land-grant universities and other institutions to explore subjects such as diseases of PLANTS and animals, crop production, PEST CONTROL, selling and uses of agricultural products, and ways to conserve agricultural resources.

How agriculture impacts the ENVIRONMENT is important. For example, researchers find ways in which farmers can increase pest control without exposing people and the environment to toxic materials and CARCINOGENS. Researchers look for ways farmers can obtain the best crop yields from available land without using up all the nutrients in the SOIL.

The USDA also helps agriculturists by compiling reports on current crop production, crop prices, farm operating costs, and world agricultural data. It also provides technical assistance to **developing nations** in an effort to improve their food production and grants loans to **farm cooperatives** to help provide telephone and electric service to rural areas.

CONCERNS FOR CONSUMERS

While it works on agricultural concerns, the USDA also helps consumers. The USDA tries to ensure, through its food assistance projects,

---

THE LANGUAGE OF THE ENVIRONMENT

developing nations also known as "Third World" nations; 120 different countries, concentrated in Africa, Asia, and Latin America, that are less advanced than industrialized, free-market countries like the United States.

farm cooperatives groups of farm owners who band together in associations that provide mutual assistance in producing and marketing their products.

---

that Americans eat adequate meals. Such projects include the Food Stamp Program, in which needy people use stamps to help pay for groceries they purchase; school lunch programs that provide free, nutritious meals for needy children; and a program for expectant mothers that furnishes foods with nutrients essential for the developing unborn child.

Through continuous nutrition education, the USDA helps consumers learn to select healthful and economical foods. Food manufacturers are now required to have labels, titled *Nutrition Facts,* on their

---

products to help consumers learn more about the foods they eat. Each label lists the nutritional value of the food inside, and any claim, like *low-cholesterol* or *fat-free,* may be used only if the food meets USDA standards for these categories. A label's nutritional information applies to adults and children over age two—younger children have different nutritional needs.

DEPARTMENT DIVISIONS

Established by Congress in 1862, the Department of Agriculture was initially headed by a commissioner. In 1889, the department head became a cabinet-level position. Department staff include the deputy secretary, or chief assistant, two under secretaries, seven assistant secretaries, an inspector general, and a general counsel.

One of the under secretaries directs small community and rural development through the Farmers Home Administration, Federal Crop Insurance Corporation, and Rural Electrification Administration. The other handles international affairs and commodity programs through the Agricultural Stabilization and Conservation Service and the Foreign Agricultural Service. Assistant secretaries oversee many agencies, including the Agricultural Marketing Service, Federal Grain Inspection Service, Food Safety and Inspection Service, Forest Service, Soil Conservation Service, Food and Nutrition Service, Human Nutrition Information Service, the Office of Operations, the National Agricultural Statistics Service, Economic Research Service, and the Agricultural Research Service.

CHANGES OVER THE YEARS

From the beginning, the goal of the USDA was to improve agricultural production and research. As modern farming procedures resulted in more abundant harvests, the department placed more emphasis on the marketing of farm products and on supporting prices. In the 1960s, the USDA began promoting the expansion of trade both at home and abroad. [*See also* AGROECOLOGY; HEALTH AND NUTRITION; and MALNUTRITION.]

Department of the Interior

▌ Executive department of the U.S. government responsible for developing and conserving the nation's NATURAL RESOURCES. The Department of the Interior is also responsible for directing programs for Native Americans and in U.S. overseas territories and possessions. The department is headed by the secretary of the interior, a member of the president's cabinet, who is appointed by the president, with Senate approval.

◆ Yellowstone National Park.

◆ The U.S. national parks are under the jurisdiction of the Department of the Interior.

◆ Yosemite National Park.

◆ The Department of the Interior supervises the operation of mines through its Office of Surface Mining, Reclamation, and Enforcement.

DEPARTMENT HISTORY

The Department of the Interior was founded in 1849. It was created as a federal custodian, to take over functions that had been distributed among other government agencies and departments. Congress gave it authority over the General Land Office, the Office of Indian Affairs, the Pension Office, and the Patent Office. Congress also gave the department authority over the Commissioner of Public Buildings, the Board of Inspectors, the Warden of the District of Columbia's Penitentiary, the U.S. Census, the accounts of some U.S. court officers, and control over America's mines.

As if that were not enough, Congress later added more responsibilities to the department. These included areas such as education, hospitals, labor, and interstate commerce. Over time, the department's function changed from housekeeper to protector of the country's natural resources. With this change, other executive departments and independent agencies were created to absorb many of the department's former duties.

DEPARTMENT DUTIES

Today, the Department of the Interior supervises approximately 500 million acres (200 million hectares) of federal land. Its job is to conserve and develop all water, FISH, WILDLIFE, and mineral resources, while trying to preserve the natural ENVIRONMENT. The department has a variety of other duties dealing with these federal lands. For example, it directs geological studies of all federal land and water resources and leases federal offshore areas to developers in search of oil, NATURAL GAS, and other MINERALS. The department also reclaims dry DESERT land in western states by planning and implementing the use of IRRIGATION systems and manages recovery and replanting programs for land that was used for STRIP MINING. The Department of the Interior is also the trustee for all Native American lands and conducts programs to assist Native Americans. Each of these major areas falls under the supervision of one of the assistant secretaries for the Department of the Interior.

DEPARTMENT DIVISIONS

Today, the staff of the Department of the Interior includes the deputy secretary of the interior, who is chief aide to the secretary and serves as acting secretary in the secretary's absence, a chief legal adviser, and six assistant secretaries. One assistant secretary clarifies department policy for the public, manages the department's administration, and helps plan its budget. A second assistant secretary supervises the NATIONAL PARK SERVICE and the U.S. FISH AND WILDLIFE SERVICE. A third assistant secretary directs the BUREAU OF RECLAMATION, the U.S. geological survey, and the Bureau of Mines. A fourth assistant secretary oversees the BUREAU OF LAND MANAGEMENT, the OFFICE OF SURFACE MINING, RECLAMATION, AND ENFORCEMENT, and the Minerals Management Service. The fifth assistant secretary is in charge of the Bureau of Indian Affairs (BIA), and the sixth assistant secretary directs the department's territorial and international programs. Most of the Department of the Interior's more than 60,000 full-time employees work in field offices. The others work in the Washington, DC, office. [*See also* MINING; NATIONAL PARKS; and WILDLIFE CONSERVATION.]

Desalinization

▷Process of removing dissolved salts from OCEAN water. The range of SALINITY of ocean water varies from about 1.8% in the North Sea to about 4.4% in more inland bodies of water, such as the Red Sea. However, the average salinity of ocean water is about 3.5%. Also called *desalination*, desalinization is very important in many drought-prone areas because it can change unusable ocean water into valuable fresh water.

The idea of removing salt from ocean water is not new. In the fourth century B.C., Aristotle described a desalinization technique used by Greek sailors. Desalinization grew in importance during the nineteenth century, when engineers searched for water supplies that would not damage the steam engines used in ships. In 1869, the first patent for a desalinization process was granted in England. That same year, the first desalinization plant was built by the British government, near the Red Sea, to supply fresh water to ships that came to port. The first large desalinization plant to supply water for commercial uses was built in Venezuela in 1930.

HOW DESALINIZATION WORKS

There are two main ways to desalinize ocean water: distillation and reverse osmosis. Distillation is the simpler and less costly method. During distillation, heat is used to **evaporate** fresh water from salt water, leaving the salts behind. The water that evaporates changes to water vapor (water in the form of gas). The water vapor is collected in tubes and cooled. As it cools, the water vapor changes back into liquid form. In reverse osmosis, ocean water is forced under high pressure through a filter. The filter traps the salt crystals but allows the water to pass through. This method purifies about 45% of the saltwater passed through the filter. The remaining, more concentrated saltwater returns to the sea. Reverse osmosis is commonly used to obtain drinking water in areas such as the Persian Gulf, where fresh water supplies are limited.

THE LANGUAGE OF THE ENVIRONMENT

droughts extended periods of weather with little rain or other forms of precipitation.

evaporate to convert liquid, such as water, into a gas. Evaporation plays a key role in the water cycle.

IMPORTANCE OF DESALINIZATION PLANTS

Today, there are approximately 7,500 desalinization plants worldwide. Most of these are located in countries near the Persian Gulf, such as Saudi Arabia and Kuwait. A few smaller desalinization plants also operate in the United States.

Desalinated water is most often used to supplement existing fresh water supplies. In areas located near oceans, such as southern California, the Persian Gulf, and other dry

◆ Three solar stills at Daytona Beach, Florida, produce about 500 gallons of salt-free water daily. They occupy an area of 3,000 square feet (270 square meters).

coastal areas, desalinated water is a dependable source of fresh water. This water can be used during emergency situations, such as those created by long **droughts**.

Desalinization plants throughout the world produce less than 1% of the world's fresh water supply. The limited use of desalinization results from its high cost in dollars and energy. For example, commercially desalinated water produced by the reverse osmosis method costs about $3 per 1,000 gallons (3,800 liters) in the United States. This cost is four to five times greater than what the average citizen of the United States pays for fresh drinking water. It is more than 100 times the price paid by farmers for IRRIGATION water. Thus, using desalinated water for farming is very costly.

The high cost of desalinated water also limits its use in the home as drinking water or in small greenhouses. Small reverse osmosis desalinization units with a few gallons-per-day capability can be purchased for home use. In areas such as the Persian Gulf, where energy is inexpensive, using desalinated water for certain purposes is quite practical. [*See also* AQUIFER; SALINIZATION; WATER, DRINKING; and WATERSHED.]

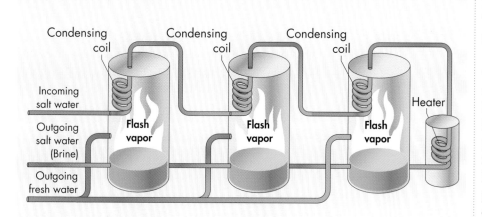

◆ Desalinization plants transform sea water into usable fresh water.

Desert

▌A land BIOME that receives less than 10 inches (25 centimeters) of rain each year. Deserts occur wherever the air above a land mass contains too little moisture to form rain or snow for most of the year. Such "dry" air forms in several ways. For example, warm air at the equator rises, cools, and drops most of its moisture at the equator itself. This air moves north or south. When the air drops again, it has little rain left to shed. This explains why most of the great deserts of the world are located in regions about 30° latitude north or south of the equator. In these regions, the effects of dry air and intense RADIATION from the sun are combined. Regions of descending dry air also occur near the North and South Poles. These regions cause "polar deserts" to form.

Other deserts are in the **rain shadows** of mountain ranges. In a rain shadow area, moist air from the sea cools and drops its rain as it passes up one side of a mountain range. As the air moves over the mountain, it has little moisture left for the land on the other side. Still other deserts form in regions where air moves over a cold sea and then onto warm land. Air that is warming up is unlikely to form rain. Some evidence also shows that deserts can form as a result of human activity.

FEATURES OF DESERTS

Any biome that receives less than 10 inches (25 centimeters) of rain each year is classified as a desert. However, not all deserts are identi-

cal. Some, such as the interior of the Sahara in North Africa, receive no PRECIPITATION at all for years at a time, and support little life. Others, such as the Sonoran Desert of North America, receive rain on a more regular basis. The Sonoran Desert supports vegetation that bursts into bloom as soon as water is available.

Deserts that have the same amount of precipitation each year may support very different PLANT and animal communities, depen-

◆ Major deserts of the world. The Sahara Desert in North Africa, the Arabian Desert in the Middle East, and the Great Australian Desert are three major deserts.

North America

Europe

Asia

Africa

South America

Australia

☐ Major deserts

ding on the annual pattern of rainfall. Water may be available for only a few days a year or over a period of weeks or months. In some deserts, precipitation may fall as snow and conditions may be too cold for plant growth. This is especially true in cold deserts of far northern or southern latitudes or in deserts at high elevations. SOIL types also influence the populations of a desert. Deserts that are covered by loose, blowing sand are very different physical ENVIRONMENTS from deserts paved with rocks. Deserts may have great differences in soil fertility as well.

Despite this variety, most deserts have certain features in common. Their lack of moisture, open landscape, and clear skies allow them to heat and cool rapidly. Thus, they generally have hot days and cold nights. In the Sahara desert of North Africa, surface temperatures as high as 185° F (85° C) are found. In the course of a year, high altitude deserts in Colorado and New Mexico may undergo temperatures ranging from -67° F to more than 100° F (-55° C to 40° C). Plants, animals, and microbes living in deserts must be adapted to rapid temperature changes and intense heat and cold. They must also be adapted to the hazards of intense solar radiation and water loss.

ORGANISMS

Desert organisms have remarkable features to help them survive in their environment. Some desert plants are short-lived **annuals** such as cheat grass, which grow and bloom only when water is available. When in bloom, they produce abundant seed

crops. Other desert plants are shrubs that drop their leaves to conserve water during periods of drought. These plants usually have long roots that reach out in search of water that seeps below the soil surface. Desert plants such as cacti, which store water in thick stems or leaves, are called **succulents.** A silvery coating of hairs protects some desert plants from the sun, while others have waxy coats to prevent water loss. Many are protected from HERBIVORES by spines. Some even repel other plants from their vicinity by secreting chemicals into the soil.

Desert animals cope with heat, cold, and drought in a variety of ways. Many avoid extreme temperatures by burrowing underground, foraging at dawn or dusk, or becoming dormant in the hottest or coldest

months. **Ectotherms** such as tortoises, lizards, and snakes regulate their body temperature by basking in the sun or resting in the shade as needed. Another ADAPTATION of many desert animals is a circulatory system that disperses heat or conserves water. A jackrabbit, for example, radiates excess heat from blood vessels in its large ears. The gemsbok, a large antelope, has a network of blood vessels in its nose that cool the blood bound for its brain. Similarly, some desert rodents have vessel networks in their nasal passages that cool air as they exhale. As a result, water vapor from their lungs collects in the nose instead of being lost from the body.

Rather than drinking, many desert rodents can form water in their bodies from the digestion of

◆ Animals like camels are adapted to desert conditions. They store water in their body and can go for days or even months without water.

◆ Deserts receive little rainfall and cannot support abundant plant life.

if the soil is fertile. However, irrigating desert land can have long-term effects on the soil that make agriculture impractical. One alternative to this risk is to seek more effective use OF SPECIES that are already adapted to the desert and do not require much water. [*See also* CLIMATE CHANGE; DESERTIFICATION; and SALINIZATION.]

carbohydrates in seeds and other plant matter. In addition, most desert animals produce highly concentrated urine to prevent water loss.

DESERT COMMUNITIES

Like other biomes, deserts support a community that includes PRODUCERS, CONSUMERS, and DECOMPOSERS. In deserts, the producers (plants) are often widely scattered or present only temporarily. Consumers of these plants include seed-eating ants, rodents, and BIRDS. Among the PREDATORS are a number of REPTILES such as lizards and snakes. Reptiles are more resistant to the dry conditions and small food supplies of the desert than are many MAMMALS and birds.

Humans are not physically adapted to true desert conditions. However, people do use deserts. Many cultures graze LIVESTOCK and harvest food in vegetated deserts or along desert margins. However, this practice has some risks. Food and water are so scarce and occur so irregularly in desert environments that the natural communities are easily disrupted. Overuse of the soil quickly degrades a desert so that it no longer supports WILDLIFE, livestock, or people.

One response to this problem has been to add water to deserts through IRRIGATION. Irrigation increases production and allows for agriculture,

Desertification

▌The process of changing a land area into a DESERT, usually as a result of human activity. A desert is a land BIOME that forms in regions that get less than 10 inches (25 centimeters)

◆ The Great Australian Desert stretches for miles through the continental interior.

of PRECIPITATION per year. Other fairly dry biomes, such as GRASSLANDS, the SAVANNA, and the **chaparral**, often lie at the edge of true deserts. In these areas, temperature changes may be almost as extreme and rain almost as scarce as in a desert. Often, just enough water falls to maintain grasses or drought-resistant shrubs.

In these marginal biomes, the role of PLANTS is critical. Their roots can reach water deep in the SOIL and bring it to the surface. This adds moisture to the ENVIRONMENT near the ground. Plant roots also help bind the soil together, making it less vulnerable to EROSION. Plant stems and leaves create shade during the day and help keep the ground from cooling off very quickly at night. As long as plants are present, a semi-arid biome has some protection against becoming as hot and dry as a desert.

CAUSES OF DESERTIFICATION

Desertification is not a new problem. Between 5000 B.C. and A.D. 200, the ancient civilizations of Sumeria, Babylonia, Assyria, Phoenicia, Egypt, Greece, and Rome existed in the Middle East, North Africa, and the Mediterranean. At that time, there was much more forest and good farmland in these regions. Today, much of that landscape is desert. Damage to the environment by humans is probably one of the reasons for this change. The Thar desert in India may also be the result of human activity, since it exists in a region where the rainfall would ordinarily be enough to keep a desert from forming.

◆ Goats survive in the desert by eating desert plants.

◆ Overgrazing of grasslands can turn a productive biome into a desert.

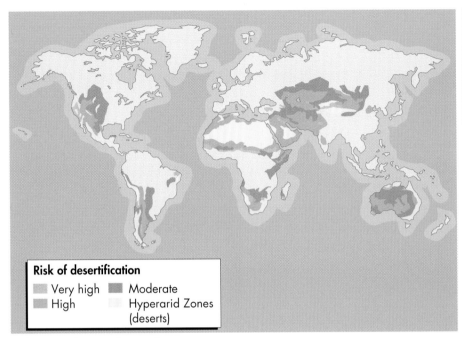

Risk of desertification

☐ Very high ■ Moderate
■ High   ☐ Hyperarid Zones
           (deserts)

◆ The map above indicates the areas in the world that are at risk of desertification.

The process of desertification usually begins with the destruction of vegetation. Though it happens most often in semiarid regions, where plant communities are easily damaged, desertification can happen even in moister areas. Natural plant cover may be destroyed by plowing for crops, or by allowing too many cows, goats, sheep, camels, or other animals to feed on the vegetation and trample the ground. Trees and shrubs may be removed for lumber or firewood. Once the plants are gone, there is much less movement of water up to the surface of the soil because there are no roots to move it. At the same time, less moisture is released into the air because there are no leaves producing water vapor. In addition, the ground is no longer shaded by vegetation. As a result, the TOPSOIL

becomes dry and loose and is subject to EROSION. The environment becomes sunnier, temperatures become more extreme, and seeds have less chance of growing in the

damaged ground. At this point, the land may remain a desert unless people make an active effort to restore the vegetation. One example of desertification in action is the southern edge of the Sahara Desert in North Africa. There, desert conditions are continuing to spread farther southward. This may be due in part to CLIMATE CHANGE, but it is probably made worse by human action.

REVERSING DESERTIFICATION

Much land that is already desertified can be restored, but **restoration** requires money, time, and at least temporarily keeping people from using the land. This can be difficult in developing countries, where poverty and overpopulation often force people into using land immediately to keep from starving. Restoration is not easy even in a developed country like the United States, if there are economic and political rewards for continuing to damage dry lands. However, some countries are taking vigorous action to restore desertified land. In China, for example, more than 14.6 million acres (5.9 million hectares) were replanted with trees and shrubs between 1978 and 1985. Further **reforestation** is still in progress. In Australia, areas have been set aside in overgrazed plains to catch seeds and regrow vegetation. Similar efforts can restore desertified environments in other areas as well. Still, preventing desertification is much less costly than repairing landscapes after they are damaged. [*See also* DEFORESTATION; DUST BOWL; OVERGRAZING; RESTORATION BIOLOGY; SALINIZATION; and SOIL CONSERVATION.]

Detergent

◆ Detergents in wastewater introduced into a stream have actually made a mass of foam on the water.

▶ An organic compound used as a cleaning agent because of its effectiveness in loosening dirt, grease, and other particles from soiled surfaces. Water molecules have a tendency to stick together. This trait gives water a greater surface tension than many other substances. All detergents contain substances called *surfactants* and *builders*.

Surfactants are chemicals that lower the surface tension in water. A lower surface tension helps loosen grease and dirt particles. Surfactants also help suspend the loosened particles in water, preventing them from being redeposited on the surface being cleaned until they are rinsed away.

Builders are chemical compounds added to control MINERALS in hard water. Hard water is water that contains many dissolved minerals. These minerals interfere with the water's ability to combine with detergents. Water can be easily identified as hard if it does not produce lather when combined with soaps. By reacting with the minerals in hard water, builders increase the cleaning power of the surfactants. Detergents may also contain brighteners, bleaches, perfumes, and coloring agents.

ENVIRONMENTAL EFFECTS

Some early surfactants used in detergents were not BIODEGRADABLE. As a result, the suds from detergents polluted waterways in areas where WASTEWATER and SEWAGE were recycled through the local water supply. In 1965, the detergent industry of the United States volunteered to change its formulas to include surfactants that are quickly broken down. Almost all detergents used today are readily biodegradable.

Another environmental problem associated with detergents is related to the use of phosphate builders. PHOSPHATES entering surface water add nutrients that are needed for plant growth. These nutrients can stimulate plant growth and may bring about an ALGAL BLOOM and accelerate EUTROPHICATION. One solution to this problem involves the treatment of wastewater to remove phosphates.

Many local and state governments have banned the sale and use of phosphate-containing detergents. In response to such bans, detergent manufacturers have tried substituting other types of builders, using extra surfactants, and adding enzymes to detergents to enhance their cleaning power. Unfortunately, some substitute builders are less effective at cleaning. In large concentrations, these builders may be toxic to PLANTS and animals. Another problem with many builders is that they greatly increase the alkalinity of water, making it irritating to the skin and eyes and weakening clothing fibers. [*See also* SEWAGE TREATMENT PLANT; WATER POLLUTION; WATER QUALITY STANDARDS; and WATER TREATMENT.]

Detritus

◆ These fungal decomposers help to recycle nutrients in ecosystems.

▌Organic waste formed from PLANTS, animals, and protists. Detritus includes the remains of dead organisms, fallen branches and leaves, and animal wastes.

Many organisms obtain their food by feeding on organic wastes. These organisms are called *detritivores*. Detritivores carry out important roles in the ENVIRONMENT. First, they prevent organic wastes from accumulating in the environment, much as street cleaners prevent GARBAGE from accumulating on city streets. Second, as they feed, detritivores help recycle nutrients through ECOSYSTEMS.

Detritivores are classified as either scavengers or DECOMPOSERS. Scavengers are animals that feed on carrion, the remains of other animals. Common scavengers in land ecosystems are hyenas, vultures,

and insect larvae such as maggots. Common scavengers in aquatic ecosystems are catfish and snails.

Decomposers are detritivores that break down complex organic substances into simpler substances, such as CARBON DIOXIDE and water, as they feed. In this way, the decomposers return the simpler substances to the ecosystem, where they can be used by plants and other organisms. In addition, many of the simple substances produced by decomposers help enrich SOIL. Two important groups of decomposers are BACTERIA and FUNGI. [*See also* CONSUMER; FOOD CHAIN; DECOMPOSITION; and HUMUS.]

Dichlorodiphenyl trichloroethane

See DDT

Dinosaurs

See MASS EXTINCTION

Dioxin

▌A general name for a large group of chemical compounds, called *heterocyclic hydrocarbons*, some of which are extremely toxic. The name is also used for one particular dioxin, TCDD (2,3,7,8-tetra-chloro-dibenzo-p-dioxin).

The dioxin TCDD is sometimes produced during the manufacture or burning of other chemicals. Tests of

its effects on laboratory animals show that it is among the most toxic chemicals ever examined and that it is a potent CARCINOGEN.

People are exposed to TCDD in various ways. Along with other dioxins, it is now widespread in the ENVIRONMENT. Because of its slow rate of DECOMPOSITION, it stays in the environment for long periods of time. Through the process of BIOACCUMULATION, it can build up in the fat of animals that people consume and in human tissues. It was also present in the HERBICIDE Agent Orange, which was extensively used in the Vietnam War. Famous cases of human exposure to TOXIC WASTE in the towns of LOVE CANAL in New

◆ Barry Krupkin speaks at a news conference about the medications he needs to take as a result of his having been exposed to dioxin in Agent Orange during the Vietnam War.

York and Times Beach in Missouri involved dioxins.

Dioxin is the subject of serious public concern. However, it is not clear exactly how dangerous it is to humans. Short-term exposure to fairly large doses can cause neurological problems, skin disorders, and other ill effects. The results of long-term exposure to low doses are less well known, and scientists have several opposing views on the subject. Dioxin has different effects on different SPECIES, so evaluating its dangers is a complicated job. [*See also* CANCER; COMPREHENSIVE ENVIRONMENTAL RESPONSE, COMPENSATION, AND LIABILITY ACT (CERCLA); DEFOLIANT; HAZARDOUS WASTE; HAZARDOUS WASTE MANAGEMENT; HERBICIDE; LOVE CANAL; PCBS; and RISK ASSESSMENT.]

Disease

See HEALTH AND DISEASE

Dissolved Oxygen

▶ The total amount of OXYGEN available in water. Oxygen makes up about 21% of all gases in the ATMOSPHERE. However, there is less oxygen in water. The amount of dissolved oxygen in water limits the number and types of aquatic organisms that can survive in the water.

The amount of oxygen that water can hold depends on a number of factors, including tempera-ture, pressure, SALINITY, and whether the water is moving or still. For instance, a quart (about 1 liter) of moving fresh water at 59° F (15° C) holds about 7 milliliters of oxygen. Thus, the oxygen content is 7 parts per million (ppm). The same amount of moving salt water at the same temperature holds about 5 milliliter or 5 ppm. As the temperature of the water increases, the amount of dissolved oxygen in the water decreases. So at 85° F (30° C), a quart (about 1 liter) of moving fresh water has only 5 milliliters of oxygen. This is one reason why FISH tend to prefer colder, deeper waters as waters warm up during summer.

Stagnant, or nonmoving, water also contributes to decreased amounts of dissolved oxygen. For instance, when water movement is slow, BACTERIA populations that feed on decaying organic matter can quickly grow in size, robbing the water of its oxygen supply. This is one reason why environmentalists are concerned when SEWAGE or fer-

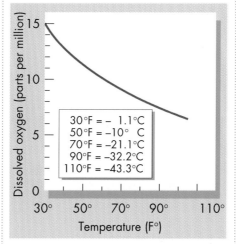

30°F = – 1.1°C
50°F = –10° C
70°F = –21.1°C
90°F = –32.2°C
110°F = –43.3°C

◆ The amount of oxygen that dissolves in water decreases as the water temperature rises.

tilizers enter a waterway. Bacteria and other organisms such as ALGAE use the nutrients in the fertilizer and sewage. The nutrients help the bacteria populations grow at a high rate, causing oxygen depletion in the water. This process, known as EUTROPHICATION, is the most common cause of fish kills, especially when waters already have less oxygen due to warm temperatures. [*See also* ABIOTIC FACTORS; ALGAL BLOOM; and OXYGEN CYCLE.]

DNA

▶ The substance in a cell that carries information for reproducing and making a new organism. Deoxyribonucleic acid (DNA) carries the genetic code for cells to make enzymes and other cell products.

SEARCHING FOR THE GENETIC CODE

The discovery of DNA was a great achievement in GENETICS. People had always observed that some features, or traits, were passed from one generation to another. A baby might "have his mother's eyes," or the seeds from a tall corn plant might grow tall corn plants. People also learned that babies, corn plants, and all other living things begin as a single cell. These observations led to questions. How was information passed from the cells of parents to the cells of their young? Where, in one cell, were the instructions for building a whole organism?

There are many kinds of molecules in a cell: sugars, starches, fats, proteins, and so on. Some molecules store energy; some hold the cell together; others help with chemical reactions. It seemed that some type of molecule had to carry a set of instructions in code, so scientists began searching for it. By 1952, Alfred D. Hershey and Martha Chase showed that the code was in a huge molecule (or *macro*molecule) called deoxyribonucleic acid, or DNA. A year later, James D. Watson and Francis Crick described the shape of this molecule.

HOW DNA WORKS

A molecule of DNA is like a long, thin rope ladder. It is usually wound up tightly in a cell. When DNA is to be used, a part of it unwinds and splits lengthwise, or "unzips," forming two half-ladders. Each half has a set of half-rungs sticking out. These are sets of instructions. Each half-rung is made of either adenine (A), thymine (T), guanine (G), or cytosine (C). Ten half-rungs on one side of an unzipped DNA molecule would read like a ten-letter code, for example "CTTGGGTTCA."

REPLICATION

What happens next depends on what the cell is doing with its DNA at the moment. It may be making a copy of the DNA molecule or it may be using the code in the DNA molecule to make proteins. Either way, substances come and attach to the half-rungs according to certain rules. For example, if the "unzipped" DNA molecule is being copied, this will happen: Adenine (A) will come and

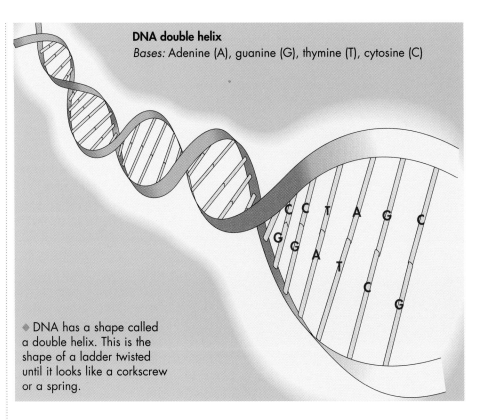

DNA double helix
Bases: Adenine (A), guanine (G), thymine (T), cytosine (C)

◆ DNA has a shape called a double helix. This is the shape of a ladder twisted until it looks like a corkscrew or a spring.

attach to any thymine (T) that is sticking out, and thymine (T) will attach to any adenine (A) that is sticking out. In the same way, guanines and cytosines (Gs and Cs) will come and attach to any cytosines and guanines sticking out. When these have all attached, the half-rungs are whole again.

A new side for the ladder is also made, from sugars and PHOSPHATES. Thus, the DNA molecule has unzipped into two halves, each half has ordered the building of a replacement half, and in the end there are two whole DNA molecules. The extra copy can then be passed on to new cells, including cells that grow into offspring.

PROTEIN SYNTHESIS

If DNA is being used to build proteins, other substances again attach to a section of "unzipped" DNA, but they do not stay there. Instead, long molecules assemble themselves by lining up one piece at a time along the DNA code. These molecules then go off to other parts of the cell and cause proteins to be made. The proteins take part in reactions that lead to growth or upkeep of the organism. Meanwhile, the halves of the DNA molecule zip back together.

A segment of DNA with the code to make a certain protein is called a GENE. The code in the example above had only ten letters, but the code of a real gene might have hundreds of letters.

While they are not the only functions of DNA, these two processes of DNA—replication and protein synthesis—are extremely important to all living things.

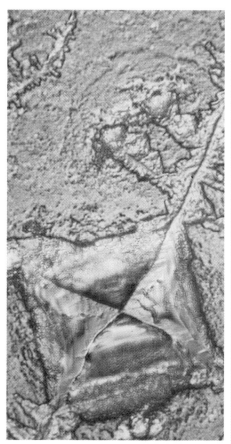

♦ Genetically engineered DNA insulin is used to treat diabetes.

Replication allows the genes of an organism to be copied for its offspring. Protein synthesis makes it possible for a large complicated organism to grow from one cell.

DANGERS FOR DNA

It is very important for DNA molecules to be copied correctly. If the codes in the molecule are changed, they may give the wrong instructions to the cell. This can make the cell abnormal, a condition known as a mutation. It may not work well, its function may change, or it may die. It may become cancerous and threaten the whole organism. If an abnormal cell is in an embryo, it may cause birth defects.

DNA is in fact constantly being damaged and changed. One cause of damage to DNA is collision with other particles moving through its ENVIRONMENT. For example, ULTRA-VIOLET RADIATION in sunshine constantly breaks the DNA of human cells by hitting the molecules with high-energy particles. Also, some chemicals set up chains of reactions in cells that lead to DNA damage. Luckily, cells have a very good system for quickly repairing damaged DNA. Thus, most of the time there is no permanent change in the genes. The repair system can be overwhelmed, however, if DNA is subjected to too much RADIATION or too many CARCINOGENS, or if a person is in poor health. In these cases, there is a greater chance of permanent damage to DNA.

While threats to DNA have always been present, some are increasing. People have always lived in sunlight, around rocks that emitted high-energy particles, and eaten PLANTS and FUNGI that contained carcinogens. However, in the last few generations, human technology has introduced many new chemicals and new sources of high-energy particles, such as RADIOACTIVE WASTES, into the environment. The amount of ultraviolet light reaching Earth's surface also seems to be increasing due to ozone depletion in the ATMOSPHERE. These changes can affect humans and other SPECIES. A worldwide decline in frog populations, for example, may be partly due to an increase in fetal damage to DNA in

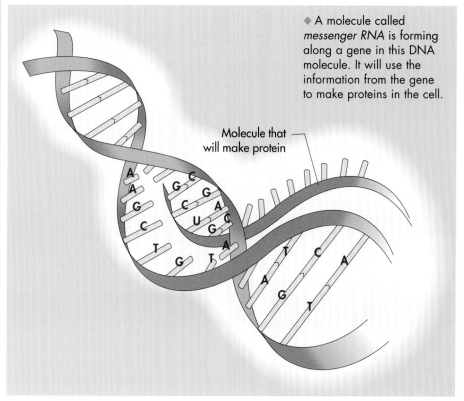

♦ A molecule called *messenger RNA* is forming along a gene in this DNA molecule. It will use the information from the gene to make proteins in the cell.

Molecule that will make protein

frog eggs that are laid in sunny ponds. It is clearly important to understand the effects of the environment on DNA in order to reduce the chance of damage. [*See also* AMPHIBIANS; RADIATION EXPOSURE; and X RAYS.]

Dolphins/Porpoises

◗ Common name for many of the smaller SPECIES of *odontoceti*, or toothed WHALES. Dolphins and porpoises are the smallest members of the family of marine MAMMALS commonly called whales. Many people consider dolphins and porpoises to be attractive, sociable, and intelligent animals. Dolphins and porpoises have become the subjects of serious scientific study and a concern of WILDLIFE preservation over the last 50 years.

The terms *dolphin* and *porpoise* are sometimes confusing. Some dolphin species have common names such as "killer whale" or "blackfish." Such names are misleading because dolphins are mammals and not FISH. There is also a "dolphin fish" that is a fish valued for its use as food.

Small dolphin species resemble their close relatives the porpoises. However, a small dolphin species can be distinguished from a porpoise species by its longer snout, cone-shaped teeth, and generally larger dorsal (back) fin. Porpoises have no protecting snout, their teeth are spade-shaped, and they have a small dorsal fin.

◆ Dolphins, such as these common dolphins (top), and porpoises, such as this harbor porpoise (bottom), are the most common types of whales.

There are dozens of species of dolphins and porpoises. They vary in size, appearance, and habits. The harbor porpoise (*Phocoena phocoena*) grows no larger than 5.5 feet (1.7 meters) long. In contrast, the largest dolphin, the orca (*Orcinus orca*), can reach a length of about 30 feet (10 meters). While most dolphins and porpoises eat fish and squid, the orca eats mammals and BIRDS as well.

The HABITAT of the porpoises varies from cold seas to the tropics. Some species, such as the orca, hunt along coasts, while others, such as the spotted and spinner dolphins, have a large hunting range in the open sea. Some species live in the large rivers of South America, India, Southeast Asia, and other regions. Amazon river dolphins actually live among the trees in flooded tropical RAIN FORESTS during part of the year.

Before the twentieth century, many dolphin and porpoise species were practically unknown. But by the 1930s and the 1940s, some animals were being kept and trained in captivity. They were extensively studied, which helped scientists learn more about them. In addition, advances in technology opened the way to studying such animals in their natural ENVIRONMENT.

ECHOLOCATION

Dolphins and porpoises rely in part on echolocation to gather information about their surroundings. The ability to echolocate was first observed in bottle-nosed dolphins (*Tursiops truncatus*) and has been studied most in this species. When dolphins and porpoises echolocate, a rapid series of clicks is made by a structure inside the animal's head. Each click is so intense that a person in the water can actually feel the pressure from the barrage of sound waves emitted by an approaching dolphin. The sound waves cause vibrations, or ringing, in the objects struck by them. These sound waves also reflect back as echoes from the objects. A dolphin perceives the echoes and other vibrations through its lower jaws and the front of its head. It then interprets the sounds to determine the size, composition, shape, and distance of nearby objects.

BEHAVIOR

Nearly all dolphins and porpoises are social animals that live in groups numbering from only a few individuals to groups of hundreds or even thousands. Many species have structured groups, often based on family relationships, with dominant and subordinate members. Within these groups, communication and cooperation are important. Group members in distress may be assisted or defended by their companions. The most social species make a variety of chirps and whistles, which seem to identify individuals and indicate emotions. Like many other mammals, dolphin and porpoise mothers invest a great deal of time and energy in their young. Often, other adults also join in caring for young.

◆ Special structures in the head of dolphins and other toothed whales allow them to echolocate. The external ears of these animals are very small. Sound is received through the jaw and other parts of the head.

Diagram labels: Vestibular sac, Blowhole, Cranium, Tubular sac, Nasal plug, Melon, Brain, Premaxillary sac, Nasal tract, Lower jaw, Acoustic window in mandible, Larynx, Tympanic bulla, Trachea

◆ Dolphins are freed from the circle of a tuna net by being allowed to escape over the top of the edge. If this practice is followed, it can save tens of thousands of dolphins each year.

INTERACTIONS WITH HUMANS

The tendency of wild dolphins and porpoises to approach humans has been recorded since ancient times. There are many stories, not all of them true, of drowning people being supported and pushed ashore by dolphins. It seems that dolphins sometimes extend the same playfully helpful behavior toward humans that they would toward other dolphins.

Much research has been and continues to be conducted on dolphins and porpoises. People studying them in nature can learn how they migrate, hunt, defend themselves, and interact in large mixed groups. People studying trained, well-kept **captive** animals learn a great deal about their social behavior and conduct tests on their ability to solve experimental problems. Research on dolphin and porpoise intelligence indicates that these animals have abilities roughly between those of dogs and the greater **primates**. Their trainability makes them valuable to marine parks as performers and valuable to the military for performing undersea tasks.

HUMAN IMPACT ON DOLPHIN POPULATIONS

The greatest impact on dolphin populations has been the drowning of millions of dolphins that become caught in nets used for catching

TUNA. In the 1960s and 1970s, tuna fishers killed more than 100,000 dolphins in the eastern tropical Pacific each year. Strong public protest led to changes in the ways tuna are captured. These changes reduced the number of dolphins killed to about 14,000 dolphins a year in the 1970s. However, the killing increased dramatically in the 1980s, as new countries began tuna fishing in the eastern Pacific. By the late 1980s, some international agreements had been reached on restricting and monitoring the use of large nets in tuna fishing.

One indirect human influence on dolphins and porpoises is the discharge of pollutants into their habitats. The effects are hard to measure. However, pollutants are suspected of harming the health of dolphins and porpoises and of increasing their chances of developing disease or dying. [*See also* ENDANGERED SPECIES; FISHING, COMMERCIAL; MARINE MAMMAL PROTECTION ACT; and OCEAN DUMPING.]

Douglas, Marjory Stoneman (1890–)

▌Environmentalist and author, often called the "Grandmother of the Glades" because of her work to preserve the Florida Everglades. Marjory Stoneman Douglas was raised in New England. She moved to Florida at age 25 and became a reporter for the *Miami Herald.* As a journalist, she wrote about people and places in Florida, especially the unique Everglades.

In the 1920s, Douglas was against a plan to drain the Everglades. Many people thought of the region as useless swampland. They wanted to drain the land to create space for the building of farms and homes. Douglas wrote that the WILDERNESS area should be preserved. She stressed that the Glades was not a useless swamp, but flowing water that she called a "river of grass."

For almost 20 years, Douglas wrote about the Everglades and worked with a committee to convince the government to make the Everglades a national park. EVERGLADES NATIONAL PARK was officially opened in 1947. This was the same year that Douglas's book, *The Everglades: River of Grass,* was published.

After the park opened, Douglas wrote about nearby developments that threatened the Glades's ECOSYSTEM. These projects were polluting the water and destroying the wildlife community. In the 1980s and 1990s, Douglas campaigned for a cleanup of the Everglades. The cleanup was slow in coming. Later, a Florida nature center and school were named in honor of Douglas, the "voice of the Everglades." [*See also* BIOLOGICAL COMMUNITY; CARSON, RACHEL LOUISE; CONSERVATION; ECOFEMINISM; NATIONAL PARKS; and WETLANDS.]

Dredging

▌Method of removing MINERALS and SEDIMENTS from ocean floors, lake floors, and streambeds by scraping or vacuuming. Sediment makes up a large part of the WATER POLLUTION in the United States and other parts of the world. EROSION and RUNOFF from farmland, FORESTS, and urban construction sites cause about 83 billion tons (84 billion metric tons) of SOIL and sediment to appear in waterways each year. This sediment fills lakes and RESERVOIRS, obstructs shipping lanes, clogs hydroelectric turbines, and makes water purification more costly.

Dredging clears away sand and silt that accumulate at the mouths of rivers or behind DAMS. This MINING method is used to obtain sand and gravel for construction. However, some dredged materials, especially those buried near industrial sites, are contaminated by chemical pollutants. These materials can easily recontaminate the water during dredging. Opponents of dredging fear that it may disturb FOOD CHAINS by removing nutrients needed by organisms from the water. Dredging may also adversely affect sea currents, CORAL REEFS, and beaches. Another environmental problem caused by dredging is where to safely dispose of dredged materials contaminated by pollutants. [*See also* WATER POLLUTION.]

Dubos, René Jules (1901–1982)

◗ French-born **microbiologist** who was a groundbreaker in the development of antibiotics and founder of the René Dubos Center for Human Environment. In 1938, Dubos became an American citizen, and in 1939, developed the antibiotic tyrothricin from a substance produced by soil BACTERIA. It was the

◆ Dredging clears away sand and silt that accumulate at the mouths of rivers or behind dams.

first **antibiotic** to be commercially manufactured and used for fighting infections in wounds. His success paved the way for other microbiologists to develop penicillin and streptomycin.

In the 1940s, Dubos became an advocate of ECOLOGY. He became more aware of the relationships between people and the natural and social ENVIRONMENT. He wrote many books, including *So Human an Animal* (1968), which earned the 1969 Pulitzer Prize. In 1975, Dubos founded the Center for Human Environment, a nonprofit organization with experience in ENVIRONMENTAL EDUCATION. Dubos believed the center would help the general public and decision makers resolve environmental conflicts and establish environmental values. [*See also* CONSERVATION and ENVIRONMENTAL ETHICS.]

Dust Bowl

❚The southern Great Plains area of the United States that suffered from severe soil EROSION from wind which caused dust storms in the 1930s. The area of the United States where Colorado, Kansas, Oklahoma, Texas, and New Mexico share borders is sometimes called the Great Plains or the High Plains. Because it receives very little rainfall each year, the area is almost a DESERT. Millions of years ago, this region of the United States was covered with lakes and swamps. Fine silty SOIL settled out from the water. Many

years later, giant rivers of ice called GLACIERS deposited a hard rock called *caliche* in the area. WEATHERING broke the rock down into very fine particles. More water then covered the area, depositing more fine silt in the area.

The lakes that once covered the Great Plains area dried up long ago. A thick layer of fine powdery soil covered the region. High winds with speeds of 15 to 30 miles (24 to 48 kilometers) per hour blew down the eastern slopes of the Rocky Mountains onto the Plains. When the fine soil there was blown into the air as high as 15,000 feet (4,600 meters), it created a dust storm. Dust storms have been a feature of the Great Plains for hundreds of years. The sand hills of Nebraska to the north and of Arkansas to the south are evidence of this.

Human activity in the Great Plains made the dust storms worse. In the late nineteenth century, cattle ranchers moved into the area. The ranchers allowed their cattle to graze on the short grasses that grew naturally in the area. Despite the dust storms and snusters—snow and dust storms—the ranchers stayed. Over time, more people moved to the Great Plains. The railroad came in. Oil and gas were discovered in the region. Then in 1905, a nearly 50-year-long drought ended.

Farmers moved to the area and began raising crops. To encourage farming, the federal government sponsored training to teach special techniques for farming the Plains. H.W. Campbell was considered the father of "scientific dryland" farming. According to him, the real difficulty in the semiarid belt is loss of

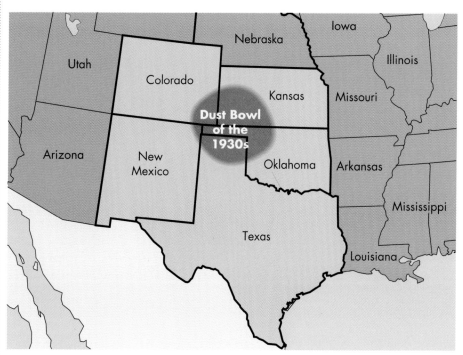

◆ The Dust Bowl of the 1930s affected Colorado, Kansas, Oklahoma, Texas, and New Mexico.

◆ Dust storms remove topsoil and transform grasslands into deserts.

the sun. A deep freeze killed the winter wheat, and ice pulverized the soil of the wheat fields. Throughout the year, dust storm after dust storm pounded the area—139 in all for the year 1933.

Associated Press reporter Robert Geiger wrote, "Three little words—achingly familiar on a Western farmer's tongue—rule life today in the dust bowl of the continent—'if it rains.'" Geiger had given the area its name—Dust Bowl.

The Dust Bowl storms continued until 1940. During that time, the federal government began a campaign to halt the farming of the area and again plant the short buffalo grass that once grew there. Many farmers resisted this suggestion and developed solutions of their own. They planted cover crops to hold the soil in place and plowed furrows that could break the wind. One farmer invented a new cultivator that chiseled the soil into large windbreaking clods and created small holes for trapping moisture.

By the 1940s, the rains came. The dust storms died out. The government managed to replant 15 million acres (6 million hectares) of GRASSLANDS. These areas, now designated as NATIONAL GRASSLANDS, were developed just in time to deal with the severe drought of the 1950s. [*See also* AGROECOLOGY; BIOME; CLIMATE CHANGE; and ICE AGE.]

too much moisture by evaporation and not lack of rainfall. He believed this could be controlled by proper cultivation. Campbell encouraged farmers to keep their fields freshly plowed to provide a cover of protective lumps of soil to trap moisture. Plowing also destroyed the weeds that used precious moisture.

The farmers planted winter wheat in the fall and harvested the grain in the spring. The field was left bare of crops for the summer to save moisture. It was plowed two or three times during the summer to keep down weeds. In the 1920s, the faithful plow horse was retired in favor of mechanical plows and tractors. With these new farm machines, farmers could produce huge areas of plowed, pulverized soil.

In September 1930, the weather of the Great Plains region changed. Suddenly there were floods, winds, more floods, blizzards, and then dry summers. In January 1933, winds pulled at the fields bare of crops and plowed to powder, pushing great clouds of dust high into the air. The dust was so severe it blocked out

Dust Storms

See DUST BOWL

E

Earth Day

▶An annual celebration of Earth centered on environmental issues. The first Earth Day took place on April 22, 1970, in the United States. More than 20 million people participated in this nationwide event, which emphasizes ENVIRONMENTAL EDUCATION and activism.

To a number of people, this day marked the beginning of the modern environmental movement. Following the founding of Earth Day, national support for CONSERVATION increased greatly. Environmental legislation became a major concern. Several important environmental laws, including the CLEAN AIR ACT and the CLEAN WATER ACT, were passed or revised, and the ENVIRONMENTAL PROTECTION AGENCY (EPA) was established.

Today, Earth Day is celebrated worldwide. Festivals, protests, work- shops, and other activities high- lighted the twentieth anniversary of Earth Day in 1990. Reaching more than 200 million participants in over 140 countries, Earth Day seems to have achieved its goal of raising worldwide environmental aware- ness. [*See also* GREEN POLITICS.]

Ecofeminism

▶Environmental philosophy partly based on the principles of femi- nism. The term *ecofeminism* was first coined in 1974. Ecofeminists emphasize the connection between environmental problems and sys- tems of domination. They believe that the domination of men over women, wealthy people over poor people, and people over animals threatens the future of the planet.

Ecofeminists want to ensure a future for Earth through practices that do not abuse people or nature. An ecofeminist might work not only for the CONSERVATION of the ENVIRON- MENT, but also for the rights of INDIGENOUS PEOPLES and for ANIMAL RIGHTS.

The principles of ecofeminism are drawn from the environmental peace movements, Native Ameri- can beliefs and other sources, as well as from the feminist move- ment. [*See also* DEEP ECOLOGY.]

Ecojustice

See ENVIRONMENTAL JUSTICE

Ecological Economics

▶A field of study that examines the relationship between ECOSYSTEMS and economic systems. The goal of ecological economics is to achieve an ecologically and economically sustainable world.

Some economists are con- cerned that traditional economic concepts fall short in their ability to take into account ecological prob- lems. Research in ecological eco- nomics focuses on sustainability, assigning values to NATURAL RE- SOURCES, accounting for the costs of environmental degradation and restoration, and modeling the interplay between economic and ecological systems. Ecological economists are interested in the meaning of concepts like SUSTAIN- ABLE AGRICULTURE and the relation- ships between ECOLOGY, economics, and culture. They study ways of assigning value to natural resources and BIODIVERSITY. [*See also* COST- BENEFIT ANALYSIS; ENVIRONMENTAL ETHICS; and ENVIRONMENTAL IMPACT STATEMENT.]

Ecology

The study of the interrelationships among living things and between living things and the ENVIRONMENT. The word *ecology* was first coined in 1870 by German biologist Ernst Haeckel. The term *ecology* comes from the Greek words *oikos,* meaning "house," and *logos,* meaning "study of." Ecologists study all of the relationships and interactions of organisms on Earth, the "house" that we live in.

The modern science of ecology has no clear beginning. The natural history studies of Aristotle nearly 2,500 years ago and the **classification** work of Carolus Linnaeus and others during the seventeenth and eighteenth centuries played key roles in our understanding of nature by studying the **diversity** of life. Modern ecology emerged in the early twentieth century as scientists began to study why and how living things interact.

Ecology combines information from geology, chemistry, biology, physics, and mathematics. The main goal of ecology is to learn how nature works. However, as the environment becomes increasingly threatened by the demands of a growing human population and human activities, ecology has developed a new goal—to help scientists develop methods to protect the natural world.

ECOSYSTEMS: THE BASIC UNITS OF STUDY

The study of ecology is based on the concept of the ECOSYSTEM. An ecosystem is all living and non-living things in an area, together with their interactions. A typical pond ecosystem, for example, contains the FISH, turtles, frogs, ALGAE, waterlilies, cattails, and other SPECIES that live in or around the pond. It also includes the water of the pond, OXYGEN dissolved in the water, and sunlight that reaches and penetrates the pond water. An ecosystem can be as small as a puddle or as large as an entire FOREST or OCEAN. Most ecosystems are complex and contain hundreds to thousands of interacting species. For instance, a CORAL REEF ecosystem contains thousands of fish, shrimp, crabs, sea stars, jellyfish, sea slugs, and a vast array of other species.

These species interact with each other and with the nonliving factors of the environment in a variety of interesting and unusual ways.

Levels of Organization in Ecosystems

The study of an individual organism, such as a squirrel, might reveal what types of food the squirrel eats, how often it eats, and how far it travels to search for food. However, squirrels also interact with other squirrels and other species. To get a complete picture of an organism's way of life, ecologists must study organisms at several different levels. For instance, ecologists might study the feeding behavior of an individual squirrel, or they might focus on the interactions within the entire population of squirrels living in an area.

Scientists might also be interested in how squirrels interact with other species within the BIOLOGICAL COMMUNITY. A biological community is made up of all the interacting populations of an ecosystem. A BIOME is a region with distinctive CLIMATE, geography, and organisms. Each biome can be viewed as a large ecosystem. Scientists recognize seven major land biomes. These biomes are the TUNDRA, TAIGA, temperate forest, RAIN FOREST, GRASSLAND, SAVANNA, and DESERT. Organisms within a given biome often have similar ADAPTATIONS because they live in areas with similar environmental conditions. For instance, many species of trees that live in temperate forests on different continents share the adaptation of dropping their leaves in order to

Individual organism—a squirrel

Population—a group of squirrels

◆ Ecologists study ecosystems at various levels of organization.

Biological community—all organisms

A niche of an organism's can be thought of as its "job" within an ecosystem.

Another way of describing an organism is by placing it in a TROPHIC LEVEL. This is a term that ecologists use to distinguish where an organism fits in a FOOD CHAIN or FOOD WEB. The three basic trophic levels are those of PRODUCER, CONSUMER, or DECOMPOSER. Producers, such as PLANTS, algae, and certain species of BACTERIA, are organisms that are able to produce their own food using energy and certain raw materials from the environment. In addition, producers—which are eaten, directly or indirectly, by all other types of organisms—produce all the food and energy available in an ecosystem. Consumers are organisms that obtain their food and energy by eating other organisms. Consumers are further divided into HERBIVORES, CARNIVORES, and OMNIVORES. Herbivores are consumers that eat plants, carnivores are consumers that eat other animals, and omnivores eat both plants and animals. Examples of consumers are animals and most bacteria, FUNGI, and protists. Decomposers, such as some species of fungi, bacteria, and protists, are organisms that break down or decompose dead organic matter as they feed.

conserve water in the winter. Desert organisms have different adaptations to help them conserve water.

Habitats and Niches

Each member of a biological community has a certain place where it lives—its HABITAT. An organism's habitat is the area where it can find all the materials it needs to survive.

All organisms play important roles in the functioning of the ecosystem in which they live. Con-sider the gray squirrels that live in deciduous forests. These forests provide the squirrels with nuts, ber-ries, and seeds for food; hol-low trees for winter shelter; tree branches for traveling; loose soil for storing food underground, and water for drinking. The habitat for a plant would include the right mixture of CLIMATE, soil type, sun-light, and growing space the plant requires.

Another term that ecologists use to describe an organism's place in the ecosystem is the word NICHE.

Interactions Among Organisms

To gain a complete picture of an organism's niche, ecologists study the interactions organisms have with other species. Organisms do not exist alone. To obtain food, shelter, and the other materials they need to survive, organisms often interact with other species.

Competition and predation are two types of interactions between species. Predation is an interaction in which one animal hunts and kills another animal for food. A BALD EAGLE eating a fish it has captured is an example of predation. Another type of interaction is COMPETITION. Squirrels, for instance, compete with birds, INSECTS, and other animals for the seeds and nuts of trees. Plants compete with each other and with animals for space to grow and for sunlight.

A third type of interaction is SYMBIOSIS, a close association between two or more species. LICHENS, the dry, crusty, multicolored organisms sometimes found on rocks and tree trunks, demonstrate a symbiosis known as MUTUALISM. Lichens are actually two organisms that exist as one. They are part fungus and part alga. Both species receive benefits from this close relationship. For example, the fungus obtains a steady supply of food that is produced by the alga. The alga receives a safe, undisturbed home and the raw materials needed to carry out PHOTOSYNTHESIS.

Other types of symbiotic relationships are COMMENSALISM and PARASITISM. Commensalism is a partnership in which one species benefits without harming or helping the other species. Parasitism is a partnership in which one species lives on or in another organism. The parasite feeds on the tissues or

◆ A raccoon's habitat includes the tree in which it lives.

◆ Red foxes are found in Asia, Europe, and northern North America.

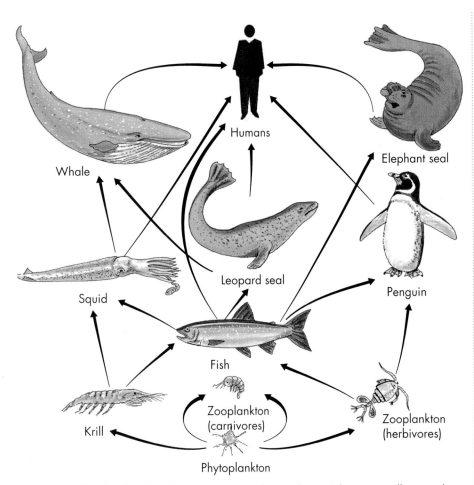

Humans

Whale

Elephant seal

Leopard seal

Penguin

Squid

Fish

Zooplankton (carnivores)

Zooplankton (herbivores)

Krill

Phytoplankton

◆ In aquatic food webs, the phytoplankton are the producers. The arrows illustrate the relationships within food chains.

drawn using arrows to indicate the direction of the flow of energy and matter from one organism to the next. A simple food chain involving humans might be shown as follows:

grass → cow → human

In this example, a cow obtains energy by eating grass. When humans eat beef, such as a hamburger, or drink a glass of milk, they obtain some of the energy that was originally produced by the grass.

Simple food chains are easy to study. However, food chains do not indicate most of the complex relationships that exist within ecosystems. To show a more complete picture of how energy moves from one organism to another, ecologists use a model known as a food web. A food web is a series of interconnected food chains. At the base of all food chains and food webs are the producers—plants, blue-green algae, or bacteria—that make food using the energy in sunlight or from chemicals in the environment. Because producers make food available to other organisms, all life is dependent on their work.

fluids of the host organism. In this relationship, one species benefits while doing harm to the other. A common example of parasitism is demonstrated by fleas living on a dog or cat.

ENERGY AND NUTRIENT CYCLING WITHIN ECOSYSTEMS

When a squirrel eats a nut from a tree, it is consuming CARBON, nitrogen, and other chemicals the tree used to produce that nut. Nuts also contain food energy stored from photosynthesis. Through the inter-

actions of organisms, matter and energy constantly cycle through the ecosystem. Ecologists study the interactions of species to make models about how nutrients and energy are transferred through the ecosystem.

Food Chains and Food Webs

All organisms need energy and other nutrients to survive and carry out their life processes. To show how matter and energy move through the ecosystem, scientists use a model known as a food chain. Food chains are typically

Nutrient Cycling

In addition to energy, organisms need nutrients to survive. Nutrients are chemical substances that organisms obtain from their foods. Nitrogen, CARBON DIOXIDE, water, and MINERALS, such as calcium and phosphorus, are chemical substances needed for the life processes of organisms. These substances also cycle through ecosystems as organisms interact with each other. Three of the major nutrient cycles that ecologists have studied deeply are

◆ Ecologists use the model of the food web to show how matter and energy move through organisms in an ecosystem.

Ecosystem

❙❙The association of PLANTS, animals, and other organisms in an ENVIRONMENT, along with the physical and chemical components of that environment. The idea of ecosystems grew from the observation that SPECIES affect one another and their environments.

THE ECOSYSTEM CONCEPT

In the early twentieth century, ecologists noted that groups of species are bound together by feeding relationships. The English ecologist Charles Elton pointed out that these feeding relationships created BIOLOGICAL COMMUNITIES.

Plant ecologist A. G. Tansley expanded on Charles Elton's idea. He noted that biological communities also interact with their physical surroundings. Biotic, or living, and ABIOTIC, or nonliving, elements, and the interrelationship among them, could therefore be studied together as a unit. Thus, someone studying a forest ecosystem would include not only species, but also SOIL, rocks, air, water, and WEATHER. A FOREST or PRAIRIE is an example of a large, land-based ecosystem called a BIOME. Within a biome, smaller units, such as marshes, ponds, or islands, are also considered ecosystems.

Ecosystems are described as "leaky" because energy and materials move from one ecosystem to another. This movement makes it

the WATER CYCLE, the CARBON CYCLE, and the NITROGEN CYCLE.

SUCCESSION: ECOSYSTEM CHANGE

Ecosystems develop and change over time in a process known as SUCCESSION. A typical succession begins when organisms known as PIONEER SPECIES begin to live on bare ground. Pioneer species are usually extremely resistant species, such as lichens and weeds, that are able to withstand extreme changes in the environment. Over time, these species change certain features of the environment, making the environment suitable for other species. Eventually, gradual changes in nutrient availability, temperature, shade, wind protection, and living space allow animals and other plant species to move into the commu-

nity. The end result of succession is a stable biological community, known as a CLIMAX COMMUNITY.

Ecological succession does not always proceed the same way in the areas that might at first appear similar. Many small factors can affect whether an abandoned field will become a pine or an oak forest after 100 years, for instance. By studying how ecosystems develop and change over time, ecologists can learn more about how ecosystems are affected by sudden natural changes like fires or hurricanes, or by human actions such as MINING, CLEAR-CUTTING of a forest, or the introduction of EXOTIC SPECIES. The science of ecology is the main way in which we gain an understanding of the consequences of environmental change, whether the result of human action or not. [*See also* BIOSPHERE and EVOLUTION.]

difficult to decide where one system ends and another begins. As a result, while ecosystems are most often studied as fairly self-contained units, all are part of a single global ecosystem.

ENERGY FLOW

The way energy flows is an important feature of an ecosystem. Energy flow can be traced through individual organisms. For example, a dandelion absorbs energy from the sun and converts some of it to plant tissue. A rabbit nibbling a dandelion absorbs energy from the plant tissue. Some of this energy is stored as animal tissue. The rest may be used in body heat or in energy for hopping. A coyote dining on rabbit absorbs some of the energy from rabbit tissue. This energy is stored in coyote tissue or used in activity and body heat. In turn, the coyote's tissue is preyed upon by parasites and then broken down after death by scavengers and DECOMPOSERS. At each feeding step, much of the energy taken in is returned to the environment as heat or undigested waste.

Energy flow can also be traced through TROPHIC LEVELS as it passes from the ecosystem's primary PRODUCERS (PLANTS) to the primary CONSUMERS (HERBIVORES), the secondary consumers (CARNIVORES), and eventually the decomposers. The flow is often quite complicated. The rate of energy flow varies with the type of ecosystem. In tropical RAIN FORESTS energy tends to be stored for a long time in the living plants. However, in SALT MARSHES much energy is stored in dead grass that floats out of the ecosystem.

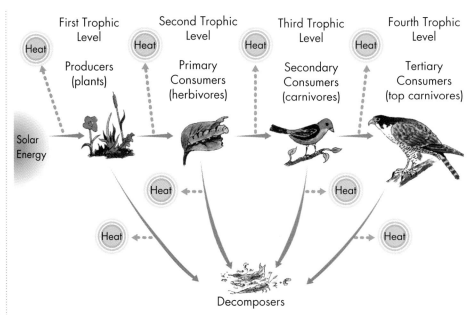

◆ Energy flow in an ecosystem. In each step, energy is lost as heat.

MATERIALS CYCLE

Another important feature of an ecosystem is the circulation of elements and compounds. Water, atmospheric gases, and soil nutrients all move back and forth between the biotic and abiotic parts of the system. The water, CARBON DIOXIDE, and atmospheric nitrogen that enter an ecosystem may be used over and over by living organisms before being lost to another system. Nutrients such as nitrogen compounds, phosphorus, sulfur, and calcium are also recycled within the system for days, years, or even centuries.

The combined effect of SOLAR ENERGY, temperature, water, and nutrients determines another important feature of an ecosystem: its primary productivity. Primary productivity is an ecosystem's rate of production of plant matter. Most healthy ecosystems have a primary

productivity that is fairly constant when averaged over many years. These systems are thus said to be balanced.

LESSONS FROM NATURE

It is important to understand how ecosystems work because many ecosystems are vital to human survival. People use grassland ecosystems for farming and GRAZING. Forest ecosystems are used for wood production. Coastal marsh ecosystems are used for seafood production. In addition, other ecosystems provide a variety of products people use. Disturbing ecosystems too much can upset their balance and reduce their productivity.

When a GRASSLAND or forest is converted to farmland, there is almost always an increase in the rate of energy flow out of the ecosystem. Energy that might have been stored in grassland root systems or forest trees instead goes into crops that are harvested and removed. A more serious matter is the increased rate of nutrient removal. Instead of being recycled, nutrients are carried off in harvested crops or washed away with eroded soil. In a similar way, harvesting lumber by CLEAR-CUTTING increases rates of nutrient and water loss from forest ecosystems. FISH from lakes and OCEANS are often harvested faster than the system can replace them. One of the many possible side effects of GLOBAL WARMING is to increase rates of water loss from terrestrial ecosystems. Because all of these processes can eventually reduce the productivity of ecosystems, it is important to take measures to prevent such problems. Good soil conservation practices, modified methods of timber harvesting, reasonable fishing quotas, and reduced carbon dioxide production are all measures that arise from understanding how ecosystems function.

Some wild ecosystems are productive without being changed for agriculture or grazing. It may seem that these ecosystems are best left unaltered to naturally produce a resource that could be harvested. However, there are some problems with this idea. For economic reasons, completely wild ecosystems tend to be overharvested to the point of collapse. Also, a wild ecosystem produces a wide variety of species. Not all of these species are immediately useful to people.

People often change the ecosystem to divert more of its energy and nutrients into human food. In addition, wild species use energy in many ways other than for growth. Domestic animals are specially bred to convert energy and nutrients into meat. The short-term return on a cow in an African grassland may well be higher than the return on a native antelope, even though the cow is less adapted to the environment.

The difficulties that arise from the way humans use ecosystems are real. The need to prevent ecosystem collapse is also very real. Efforts to understand the function of the ecosystem and find ways to maintain healthy levels of energy flow and nutrient cycling are of great value for preserving resources for both humans and other species.

◆ Coastal marsh ecosystems provide food for the young of many oceanic organisms.

Ecotourism

▌Travel designed for those who want to see and experience pristine, undisturbed ECOSYSTEMS. Many people would like to visit a pristine RAIN FOREST or the high ARCTIC. To fulfill this need, many countries—often developing nations—that have such undisturbed natural areas within their borders have turned to ecotourism. Some of the popular activities for ecotourists are bird watching, hiking, nature photography, fishing, camping, mountain climbing, and wildlife safaris.

The aim of ecotourism is to bring in needed income by providing employment to local people through business opportunities in the form of services to tourists. The idea is that if local people benefit financially from ecotourism, they are more likely to support preserving the natural area rather than exploiting its resources. Local support for conserving natural areas is crucial to their preservation.

Ecotourism has had its shortcomings, however. In many parks, for example, tourists are allowed to roam freely without guides or with guides who are not qualified. In some areas, tourists have damaged the ENVIRONMENT by disturbing or frightening the WILDLIFE. In addition, most local communities do not benefit from the income generated by the tourism. Local communities and conservation groups are often not involved in the planning of the activities. Some critics believe that involving these groups might make ecotourism more likely to conserve the area that is being visited.

◆ Tropical rain forests, such as this one in Australia, are frequent destinations for ecotourists.

Effluent

▶Discharge of WASTEWATER from a POINT SOURCE. The word *effluent* literally means "flowing out of." Geologists use the word when referring to lava flowing from a volcano or a stream flowing from a lake. In environmental science, however, the term is generally used to refer to wastewater and is used in the CLEAN WATER ACT.

The Clean Water Act is the main body of legislation that protects the coastal and inland waterways of the United States from damage caused by WATER POLLUTION. Under this act, which is implemented by the U.S. ENVIRONMENTAL PROTECTION AGENCY (EPA), industries are allowed to discharge only certain amounts and types of pollutants in their effluent. Pollutants that must be controlled include organic waste, SEDIMENT, BACTERIA, oil, and heat.

To meet the requirements of the Clean Water Act, industries, towns, and cities must obtain a POLLUTION PERMIT before they can release effluent into the ENVIRONMENT. The permit requires that a point source use the BEST AVAILABLE CONTROL TECHNOLOGY (BACT) to control POLLUTION. In addition, a permit fee must be paid before an industry, town, or city can discharge effluent legally into a sewer system or into SURFACE WATER. [*See also* INDUSTRIAL WASTE TREATMENT; MARINE POLLUTION; NATIONAL POLLUTANT DISCHARGE ELIMINATION SYSTEM (NPDES); WASTEWATER, PRIMARY, SECONDARY, AND TERTIARY TREATMENT OF; SAFE DRINKING WATER ACT; SEWAGE TREATMENT PLANT; WATER QUALITY STANDARDS; and WATER TREATMENT.]

◆ Effluent from many point sources flows through pipes to a treatment plant.

Electricity

▶Form of energy that results from the flow of electrons. Electricity is one of the most versatile forms of energy. Society has depended on electricity since it was discovered more than 150 years ago. Electricity is used to power many of the devices people use each day. Such devices include lights, stereos, appliances, telephones, industrial machines, boats, and trains. Much of the electricity that powers such devices is produced through the conversion of other forms of energy, such as mechanical, heat, and chemical energy.

SOURCES OF ELECTRICITY

Where does the electricity people use in their homes and businesses come from? Most of the electricity people use is produced at large electrical power plants. These power plants are equipped with generators that convert mechanical energy into electricity. To accomplish this, commercial power plants must use other forms of energy to run the generators.

Fossil Fuels

Anything that can turn the coils of a generator—pressurized steam or flowing water—can be used to produce electricity. However, most electrical power is generated by plants that use FOSSIL FUELS. When fossil fuels are burned, they release heat. The heat is used to boil water and produce high pressure steam. The steam is then directed against the blades of a turbine—a large, fanlike device—which sets the turbine into motion.

More than 85% of the energy in the United States comes from fossil fuels, which are used to make electricity. However, there are many drawbacks to using them. The main problem with fossil fuels is that they are extremely harmful to the ENVIRONMENT when burned. Another drawback is that these natural resources are NONRENEWABLE RESOURCES, or materials that cannot be replenished when used up.

Nuclear Power

NUCLEAR POWER is used to generate about 7% of the electricity used in the United States today. Nuclear fuels, such as URANIUM and PLUTONIUM, are less polluting than fossil fuels and can produce much larger amounts of energy than the same amounts of fossil fuels. The process of electricity generation at a nuclear power plant is very similar to the way electricity is produced at conventional plants that use fossil fuels. However, high costs and public concern about the safety of nuclear power plants and RADIOACTIVE WASTE disposal have limited the use of this energy source. These concerns have led the United States to stop building new nuclear power plants.

Hydroelectric Power

HYDROELECTRIC POWER accounts for nearly 7% of electricity production in the United States. This form of energy is clean, renewable, and leaves no waste. However, hydroelectric power plants require the construction of DAMS, which can

◆ Hydroelectric generators are responsible for about 7% of the electricity produced in the United States.

damage river ECOSYSTEMS. Such construction can be very costly. In addition, the use of hydroelectric power plants is limited to areas located near rivers and streams.

ALTERNATIVE SOURCES OF ENERGY

The problems associated with the generation of electricity from conventional sources have led scientists to look for other ways to generate electricity. Together, these methods of generating electricity are called ALTERNATIVE ENERGY SOURCES.

Solar Energy

SOLAR ENERGY is the energy derived from sunlight. The sun may be the ultimate source of energy. The use of solar energy to generate electricity is growing rapidly. Today, solar energy is used to power everything from calculators to small cities.

Solar energy can be converted directly into electricity using PHOTOVOLTAIC CELLS—devices made of silicon and other chemicals that conduct electricity when struck by sunlight. Photovoltaic, or solar, cells supply electricity for calculators, cars, and satellites. However, at present, they are very expensive and difficult to produce.

Another promising use of solar energy is the solar-thermal power plant. This power plant uses large mirrors to direct the sunlight onto a series of oil- and water-filled pipes. The heated water is then used to turn the turbine blades of the generator. A working solar-thermal power plant in California's Mojave Desert has shown that this method can supply enough electricity to run a city the size of Washington, DC.

Many scientists believe that solar energy will be the answer to the world's future electricity needs. The U.S. Department of Energy is testing these arrays at Golden, Colorado.

Wind Power

The wind is a tremendous source of energy that is both clean and renewable. This energy can be captured by a turbine and used to produce electricity when connected to an electrical generator. WIND POWER is being used in California, where more than 1,300 megawatts of electricity are generated by 15,000 wind-driven turbines. Currently, wind power supplies about 1% of California's electricity needs. This is enough to supply the electrical needs of the city of San Francisco.

Geothermal Energy

Geothermal power plants use the heat within Earth to make electricity. In this method, water is injected through long pipes that extend deep into the Earth. The hot rocks below the surface heat the water to produce steam. The steam is then directed to a turbine, which turns the generator.

GEOTHERMAL ENERGY is clean; however, it is nonrenewable. Natural processes replace the heat very slowly. For instance, geothermal energy was once a major source of electricity in New Zealand. However, excessive use of this resource depleted it.

MEETING THE ELECTRICITY NEEDS OF THE FUTURE

Since World War II, electricity consumption has increased rapidly in all parts of the world. Experts agree that with the increasing use of computers and other electronic technology, as well as the growth of the human population, this trend will likely continue.

Today, fossil fuels supply most of the world's electricity. However, as their supplies dwindle, they will become much less important. To meet the electricity demands of future generations, environmentalists suggest that people try to conserve electricity, as well as develop efficient new ways to produce it without damaging the environment.

Many scientists believe solar energy will be the answer to the world's future electricity needs. The costs of using solar energy to make electricity are high. However, technological advances will make this energy source much more cost-effective in the future. Wind power, TIDAL ENERGY, and OCEAN THERMAL ENERGY have also shown success as methods of producing electricity. These energy sources will likely be important in the future. All depend on renewable forms of energy and are relatively harmless for the environment. [*See also* ATOMIC ENERGY COMMISSION; BREEDER REACTOR; CHERNOBYL; ELECTROMAGNETIC FIELD; ENERGY EFFICIENCY; FUEL; GLOBAL WARMING; NUCLEAR FISSION; and NUCLEAR FUSION.]

Some Things You Can Do to Save Energy

1. Drive less: make fewer trips, use telecommunications and mail instead of going places in person.
2. Use public transportation; walk or ride a bicycle.
3. Use stairs instead of elevators.
4. Join a car pool or drive a smaller, more efficient car; reduce speeds.
5. Insulate your house or add more insulation to the existing amount.
6. Turn thermostats down in the winter and up in the summer.
7. Weatherstrip and caulk around windows and doors.
8. Add storm windows or plastic sheets over windows.
9. Create a windbreak on the north side of your house; plant deciduous trees or vines on the south side.
10. During the winter, close windows and drapes at night; during summer days, close windows and drapes if using air conditioner.
11. Turn off lights, television sets, and computers when not in use.
12. Stop faucet leaks, especially hot water.
13. Take shorter, cooler showers; install water-saving faucets and showerheads.
14. Recycle glass, metals, and paper; compost organic wastes.
15. Eat locally grown food in season.
16. Buy locally made, long-lasting materials.

◆ This electrical substation is located right next to a residence.

◆ Many studies have been done to look for any link between electromagnetic fields surrounding high-voltage power lines and reported cancers in children living beneath them.

Electromagnetic Field

▌Invisible electric and/or magnetic force that surrounds any device through which ELECTRICITY flows. Electromagnetic fields were discovered by the Danish physicist Hans Christian Oersted in 1819.

Oersted observed that a force field developed around a wire through which electricity flowed. When he stopped the electric current, the force field surrounding the wire disappeared. Oersted also observed that the intensity of the electromagnetic field changed, depending on how much electricity flowed through the wire. For example, a weak flow created a weak, sometimes blurred field; a strong flow

produced a strong field. Oersted's discovery made it possible for people to ring doorbells, control electric lights, and accomplish many other mechanical jobs by simply flipping a switch between OFF and ON.

Strong electromagnetic fields exist around high-voltage power lines that carry electricity from power plants to homes and businesses. Electromagnetic fields also exist around household appliances, computers, office equipment, and industrial equipment that runs on electricity. Since the 1970s, many people have become concerned that extended exposure to electromagnetic fields might be harmful to the health of humans and animals. Most concern centers on the dangers of electromagnetic fields from high-voltage power lines to people who live near them.

More than 90 studies have been published examining the link between electromagnetic fields and certain kinds of CANCERS. For example, it was believed that electromagnetic fields might be responsible for the reported cancers in people living near electric facilities and for the reported cases of leukemia and brain tumors in electrical workers. To date, there is insufficient data to determine whether electromagnetic fields pose any threats to health, but scientists continue to investigate the question. [See also CARCINOGEN and RADIATION EXPOSURE.]

Electromagnetic Spectrum

▶Range of electromagnetic RADIATION in nature. Electromagnetic radiation is classified according to its wavelength. In order, from lowest energy wavelengths to highest, electromagnetic radiation includes radio waves, infrared light, visible light, ultraviolet light, X RAYS, and gamma rays. These waves transmit sound, light, and heat through air and matter by vibrating electrical and magnetic fields at different speeds. The more energy present in the vibration, the shorter the wavelength.

PARTS OF THE SPECTRUM

The highest energy waves are gamma rays. Their wavelengths are shorter than one nanometer, or one-billionth of a meter. The lowest energy waves are radio waves. These may be many miles long and are sensed by the hearing organs of organisms. Radio astronomers study radio wavelengths that bounce off objects in space and listen for radio "background noise" in our universe.

Infrared or heat wavelengths are given off by hot objects. These waves are used in the military for detection of missiles and for night vision. Infrared rays are located in the electromagnetic spectrum just beyond the red band in visible light.

Visible wavelengths fall in a narrow range. Their wavelengths are between 350 and 750 nanometers. Living things absorb and use visible light with different wavelengths. White light is made up of all the colors of visible light. A prism can bend white light into its component colors. Each color has a different wavelength. The rainbow, or wavelengths of light making up the color spectrum, always appear in the same order according to wavelength. These colors are red,

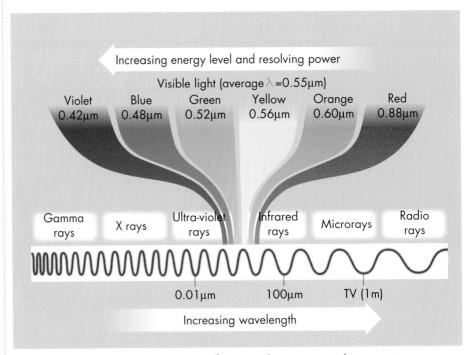

◆ The electromagnetic spectrum ranges from very low energy radio waves to very energetic gamma rays. Only part of those waves (350–750 nanometers long) are visible in white light.

orange, yellow, green, blue, indigo, and violet.

Ultraviolet light consists of the waves of light that lie just beyond the violet band in the visible spectrum. Ultraviolet light has wavelengths between 380 and 1 nanometer. Although it makes up 5% of the sun's energy, most ULTRAVIOLET RADIATION is prevented from reaching Earth by the OZONE LAYER. This is important because too much exposure to ultraviolet light is harmful to living things. On Earth, ultraviolet rays are used in fluorescent lighting and in sterilizing lamps.

Energetic X rays and gamma rays have wavelengths shorter than one nanometer. X rays are used in medicine as a diagnostic tool that allows doctors to see the bones inside the body. They are used in industry to view package contents for security purposes. Radiation from gamma rays can be used to destroy living tissues and cells. Such use is common in cancer treatment.

ENVIRONMENTAL PROBLEMS ASSOCIATED WITH THE ELECTROMAGNETIC SPECTRUM

Ultraviolet rays and gamma rays can create environmental problems. Earth's important ozone layer, which protects Earth from the sun's ultraviolet radiation, has been thinned. Chemicals such as chlorofluorocarbons, (CFCS) that are used in some AEROSOLS are believed to have caused a hole in the ozone layer. As a result, more of the sun's ultraviolet radiations can reach Earth. Production of CFCs and other ozone-destroying chemicals is scheduled to stop by the twenty-first century. However, scientists fear CFCs already in the ATMOSPHERE could continue to do damage for many years.

The use of NUCLEAR POWER poses the threat of emissions of RADIOACTIVITY into the ENVIRONMENT. Radioactive material is harmful to living cells. The release of radioactivity in the environment in large amounts could contaminate air, PLANTS, and animals in ECOSYSTEMS. Such radioactivity can be carried by the wind. Some environmentalists want the use of nuclear energy stopped until experts can guarantee this will not happen and safe storage and disposal of radioactive material can be arranged. [*See also* ELECTROMAGNETIC FIELD; GREENHOUSE EFFECT; and OZONE LAYER.]

Elephant

▌Either of two SPECIES of MAMMALS that have a trunk used for collecting food, drinking, and bathing. Elephants are the largest land mammals. The two species of elephants are the African elephant and the Indian elephant. The African elephant lives in the regions of Africa south of the Sahara DESERT. The Indian elephant lives in parts of India and Southeast Asia. A newborn African elephant calf weighs between 255 and 320 pounds (116 and 145 kilograms). In contrast, an Indian elephant weighs about 220 pounds (100 kilograms) at birth. As elephants grow to adulthood, the males grow larger in size than the females. The largest elephant

◆ Poachers kill elephants for their ivory tusks, which are used to make jewelry and carvings.

known, an African elephant named Jumbo, lived in the London Zoo during the 1800s. This elephant weighed more than 4,500 pounds (2,000 kilograms) and was almost 11 feet (3.4 meters) tall.

ELEPHANTS IN DANGER

The African and Indian elephants are the only surviving species of the mammal group *proboscidean*. The mammal group consisted of 350 species of long-snouted animals that included mastodons and mammoths. The great size of the elephant has protected it from almost all PREDATORS, except humans. Years ago, elephants traveling in herds protected themselves from hunters by closing ranks or charging. Today, however, entire elephant families are killed by poachers, or illegal hunters. The poachers kill the elephants to obtain their tusks, which are made of ivory that is then used for making jewelry and carvings.

◆ The African elephant on the left is instantly recognizable because of its large ears and the two knobs on its trunk. The Asian elephant on the right has small ears and one knob on its trunk.

In 1979, wildlife experts estimated that 1,500,000 wild elephants lived in Africa. That number began to rapidly decline. In the late 1980s, a public campaign to save the African elephant began. In 1989, the CONVENTION ON INTERNATIONAL TRADE IN ENDANGERED SPECIES OF WILD FAUNA AND FLORA (CITES) banned all trade in ivory and other elephant products. However, the ban has been difficult to enforce, and some countries want the ivory ban to be lifted because they believe their elephant population is growing. By the early 1990s, only about 600,000 African elephants remained in the wild. If the decline continues, experts believe that elephants may be extinct before the year 2000.

Both African elephants and Indian elephants are ENDANGERED SPECIES. Experts agree that the elephant needs human protection to survive. Besides poachers who kill elephants, settlers have destroyed much of the animals' HABITAT by clearing land for farms and villages. African and Asian nations have set aside land for elephant use in NATIONAL PARKS or in reserves. However, experts fear that not enough land has been provided to save many wild elephants. New laws forbid the HUNTING of elephants in parks and reserves and limit the number of elephants hunters can

◆ The number of African elephants has declined. The 1979 estimated number was 1,300,000, which declined to 600,000 in the early 1990s.

kill outside those areas. But the laws are difficult to enforce, and poachers still kill thousands of elephants per year. [*See also* EN-DANGERED SPECIES; EXTINCTION; and POACHING.]

El Niño

▌An unusually warm OCEAN CUR-RENT off the west coast of South America, especially near Peru. El Niño can have damaging effects on the marine ECOSYSTEMS where it occurs. It also appears to contribute to severe climate events worldwide, such as floods, heat waves, and droughts.

El Niño, Spanish for "the boy," was first documented in 1726, when Peruvian fishermen noticed schools of anchovy FISH and other marine organisms dying in the waters off the coast. The term refers to the Christ Child because the effects of the El Niño current are usually presented just before Christmas, during December. El Niño develops when large masses of warm, nutrient-poor, tropical water, brought by strong Pacific currents, replace the normally cool waters off Peru and Ecuador. In most years, the warming of the water is moderate and lasts for only a few weeks. However, during a severe El Niño, these warm waters are poor in nutrients and OXYGEN, and large numbers of marine organisms die. BACTERIA populations feeding on the decaying materials in the water

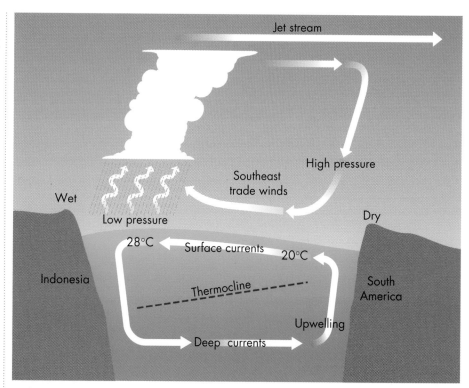

◆ El Niño appears to be linked to many unusual weather events worldwide.

explode in size, further depleting the water of oxygen.

When a large El Niño occurs, it can temporarily disrupt the economies of South American countries that rely heavily on commercial fishing. However, scientists are more concerned about the CLIMATE CHANGES caused by El Niño. For instance, scientists believe that the 1986–1987 El Niño was responsible for the record global warmth in 1987. Average temperatures that year were the warmest in over 100 years. More recently, the 1991–1992 El Niño has been linked to a number of unusual weather events around the world, including North America's extremely mild winter, severe droughts in central and southern Africa, and record flooding in South America.

Scientists are uncertain about what atmospheric or oceanic events trigger El Niño. However, they do understand that El Niño occurs in cycles and is related to changes in the global circulation patterns of the OCEANS and ATMOSPHERE. Some scientists believe that underwater volcanoes in the South Pacific may have an effect on ocean warming, thus contributing to the occurrence of El Niño. [*See also* EUTROPHICA-TION and VOLCANISM.]

EMF

See ELECTROMAGNETIC FIELD

Endangered Species

Any PLANTS or animals having so few individual survivors that EXTINCTION is probable if nothing changes. Threatened species are SPECIES that are likely to become endangered in the near future if protection is not provided. Under the ENDANGERED SPECIES ACT of 1973, a species is endangered if its survival is threatened by:

1. destruction of its HABITAT, or the area in which a species lives;

2. commercial HUNTING;

3. pollution;

4. any other natural or human activity that places it in danger.

HABITAT DESTRUCTION

The greatest threat to species is HABITAT LOSS and destruction caused by human activities. Statistics show that habitat loss causes almost 75% of the extinctions now occurring. As human populations grow, more land is needed for farms, roads, cities, and factories. As a result, habitats are often destroyed, so that some species lose the resources they need to survive, such as food, water, and shelter.

One large animal in the United States that has been harmed by habitat loss is the Florida panther. Two hundred years ago, this subspecies of the mountain lion was common in Florida. Habitat destruction, hunting, accidents with auto-

◆ Panther cubs are covered with dark spots that disappear as they grow older.

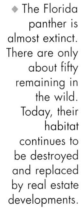

◆ The Florida panther is almost extinct. There are only about fifty remaining in the wild. Today, their habitat continues to be destroyed and replaced by real estate developments.

mobiles, and BIOACCUMULATION of MERCURY have reduced the population to somewhere between 30 and 100 animals living only in South Florida. Habitat loss was the main factor in reducing the population of Florida panthers, but now that the population is small, the other factors listed have become serious problems.

HUNTING

Unregulated commercial hunting is another human activity that threatens species. On a worldwide basis, the killing of animals for profit from the sale of furs or other animal parts endangers a number of large animal species. Many large cats, such as jaguars, cheetahs, TIGERS, and snow

Species	Found In
American bison	Central and Western United States
Spotted salamander	Eastern United States
Ivory-billed woodpecker	Southeastern United States
Black-footed ferret	Central United States
American alligator	Southeastern United States
Whooping crane	Southern United States
Florida manatee	Southeastern United States
White bladderpod plant	Texas
Golden-cheeked warbler	Texas
Green pitcher plant	Southeastern United States
Stellar sea lion	Northwestern United States
Florida panther	Southeastern United States
California condor	California
Grizzly bear	Western United States
Gila trout	Southwestern United States
Source: World Wildlife Fund 1990	

◆ Some species at risk in the United States are listed above.

leopards, are hunted commercially for their colorful furs. ELEPHANTS are killed for their valuable ivory tusks. From 1979 to 1989, the African elephant population dropped from about 1.5 million to 700,000 as a result of commercial hunting. However, in 1989, the CONVENTION ON INTERNATIONAL TRADE IN ENDANGERED SPECIES OF WILD FAUNA AND FLORA, (CITES), proposed a ban on all sales, imports, and exports of ivory. After this ban, the price of ivory fell sharply and elephant POACHING declined dramatically.

Commercial hunting in the United States was an important factor in the extinction of the American PASSENGER PIGEON and the near-extinction of the bison (also known as the American buffalo). Before European settlers came to America, an estimated 60 million to 225 million buffalo roamed the plains and PRAIRIES of the central United States. Between 1850 and 1906, growth and expansion of the U.S. population reduced the once immense buffalo population to about 300. Today, laws protect the remaining buffalo and their population has increased to about 130,000.

Many people have the mistaken belief that all types of hunting threaten WILDLIFE. Unregulated hunting can certainly place many species in danger of extinction. However, legal hunting, such as the deer and duck hunting that occurs at various times of the year, is an effective way to manage wildlife populations.

POLLUTION

POLLUTION is harmful not only to humans; it can have devastating effects on other species as well. The Florida panther mentioned earlier is threatened not only by habitat loss, but also by mercury pollution in some of the areas where it still survives. In the late 1980s, some of the last Florida panthers in the Everglades National Park were found to have died of poisoning from mercury pollution. While pollution itself is probably rarely the main cause of a species' extinction, it can be a major contributing cause, especially when other factors have greatly reduced a population.

EXOTIC SPECIES

A species can also be threatened by other species, especially when human acivities bring an EXOTIC SPECIES into a new area. For instance, the European zebra mussel was accidentally introduced into the United States and Canada by ships that traveled from Europe to the GREAT LAKES. The zebra mussel has no natural enemies in North America, allowing it to have a strong edge in COMPETITION with native freshwater mussel species, many of which are already on the endangered species list due to habitat destruction. On oceanic islands, the problem of exotic species is often severe because island species often evolve into relatively defenseless organisms. For example, in New Zealand several flightless bird species have been driven to extinction or to near extinction by the

◆ The American buffalo roamed the plains in great numbers before European settlers occupied the West. Their numbers dropped from an estimated 60 to 125 million before 1850 to 300 in 1906.

◆ The environmental conservation organization, Greenpeace, attempted to stop a whaling ship to bring attention to the killing of endangered species of whales.

◆ Sea turtles are an endangered species. Their eggs, laid in holes in the sand, are especially vulnerable to destruction.

introduction of exotic animals such as rats, cats, foxes, and other predators, which find the flightless birds easy to catch. On the island of Guam, the accidental introduction of a tree snake has resulted in several bird species in Guam to become extinct in the wild, living now only in zoos. Unless the tree snakes are eliminated, these bird species may never again exist in the wild.

PROTECTING
ENDANGERED SPECIES

As of 1994, 722 plant and animal species in the United States were listed as endangered, and 198 were listed as threatened. Most countries have laws and regulations designed to protect species. In 1973, the U.S. Congress passed the Endangered Species Act, one of the toughest environmental laws to date. Under the main provisions of this law, endangered and threatened species are protected from unlawful hunting, killing, and habitat destruction.

Several organizations work to protect endangered species internationally. Leading world efforts to preserve species is the INTERNATIONAL UNION FOR THE CONSERVATION OF NATURE (IUCN), a collaboration by several governments and private organizations. It publishes the Red Data Books, which list all species in danger of extinction around the world. An offshoot of the IUCN is CITES. CITES became well known in the 1970s and 1980s for its efforts to protect the African elephant.

In 1992, world leaders met in Rio de Janeiro, Brazil, for the UNITED NATIONS EARTH SUMMIT, a meeting organized to discuss worldwide environmental issues. One of the agreements to come out of the conference was the Biodiversity Treaty, which encourages developed countries to donate money to poorer countries for the protection of endangered species. [*See also* BIODIVERSITY; CALIFORNIA CONDOR; GRIZZLY BEAR; IVORY-BILLED WOODPECKER; LAW, ENVIRONMENTAL; SEALS AND SEA LIONS; and WATER POLLUTION.]

Endangered Species Act

Legislation enacted in 1973 by the U.S. Congress to protect threatened and ENDANGERED SPECIES from EXTINCTION. The Endangered Species Act of 1973 is one of the toughest environmental laws developed by any country. The main provisions of this legislation are:

1. The U.S. FISH AND WILDLIFE SERVICE must keep an up-to-date list of all endangered and threatened species.

2. Endangered or threatened animal species may not be captured or killed. Plant species on federal properties may not be uprooted or disturbed. No part of any endangered or threatened species may be sold or traded.

3. The federal government may not carry out any project that jeopardizes the survival of an endangered or threatened species.

4. The U.S. Fish and Wildlife Service must develop a species-recovery plan for each endangered and threatened species on the list.

◆ In an effort to raise chicks of the endangered whooping crane in captivity, George Archibald, director of the International Crane Foundation, waits for the female to lay an egg.

MAINTAINING A LIST OF ENDANGERED AND THREATENED SPECIES

The first main provision of this law states that two government agencies, the U.S. Fish and Wildlife Service and the National Marine Fisheries Service, are authorized to identify and prepare a list of all endangered and threatened species. Any decisions to add or remove a species from this list must be based on biological, rather than economic, reasons. Today, there are more than 900 endangered and threatened species on this list. Waiting to be classified are nearly 4,000 other officially recognized species that have not yet undergone a full review.

PROHIBITING KILLING OR CAPTURE

Under the second main provision of this law, it is illegal to hurt, capture, or kill any endangered or threatened animal species or uproot any plant species growing on federal land. This provision also prohibits interstate and international trade of endangered and threatened species and any of their products, such as skins, furs, or feathers.

PROHIBITING HARMFUL FEDERAL ACTIONS

According to the third main provision of this law, once a species is listed as endangered or threatened, no branch of the federal government may carry out a project that places the species at risk. In 1975, this provision of the Endangered Species Act was put to the test. At

◆ The tiny snail darter, an endangered species, held up the construction of a dam in Tennessee. Conservationists sued the federal government in an effort to stop destruction of its habitat, claiming it violated the Endangered Species Act.

that time, conservationists filed a lawsuit against the federal government to halt construction of the Tellico Dam on the Little Tennessee River. They argued that construction of the DAM would destroy the habitat of a tiny FISH called the snail darter. Because snail darters had been identified as an endangered species, opponents of the dam claimed that its construction was a violation of the Endangered Species Act. The dam was constructed, and its impact on the snail darters is being studied.

Eventually, the snail darter controversy led to a congressional amendment of the Endangered Species Act. In 1988, lawmakers agreed that governmental projects should continue if the economic benefits of a project outweigh the potential environmental effects. In 1995, the debate between the needs of the U.S. economy and the ENVIRONMENT continued. Many lawmakers proposed further amendments to the Endangered Species Act, arguing that protecting certain species places limits on economic growth. However, to date, there

have not been any serious proposals for drastically changing the Endangered Species Act.

DEVELOPING SPECIES-RECOVERY PLANS

The fourth provision of the law states that the U.S. Fish and Wildlife Service must develop individualized species-recovery plans for each endangered or threatened species on the list. A species-recovery plan is a series of steps to be taken to prevent extinction and ensure survival of a species. In some cases, species are placed in ZOOS, aquariums, or botanical gardens as the last effort to prevent extinction. In other cases, captive breeding programs are established to increase the population of an endangered or threatened species. A well-known example of a successful captive breeding program involves the CALIFORNIA CONDOR. In 1986, conservationists found that there were only nine California condors left in the wild. A captive breeding program was quickly developed and now the population of California condors is slowly recovering.

Simply putting a few individuals into zoos, aquariums, and botanical gardens does very little to preserve a species. Many animals will not breed in captivity, and they are also more vulnerable to infectious diseases and genetic disorders. Scientists have discovered that the most effective way to preserve a species is to protect its habitat. [*See also* CAPTIVE PROPAGATION and CONVENTION ON INTERNATIONAL TRADE IN ENDANGERED SPECIES OF WILD FAUNA, AND FLORA (CITES).]

Energy, Nonrenewable

See NONRENEWABLE RESOURCES

Energy Efficiency

❙❙The amount of work a machine produces compared to the amount of effort or energy needed to make it work. For example, an electric motor for a water pump produces mechanical power by using only three-fourths of the ELECTRICITY needed to run it—the other one-fourth produces heat from **friction** in the wires and the motor itself. The energy efficiency of that motor is listed as 75%. The motor's energy efficiency could be improved by reducing friction and channeling the electricity to produce more power.

IMPROVING EFFICIENCY

Engineers and scientists are conducting research to find more effective ways to **reclaim**, transport, and use our dwindling sources of FOSSIL FUELS to make them last longer. Auto engines have become more energy efficient over the past 30 years, but they still generate only about 20% of the energy that is available from the gasoline that runs them. The other 80% is wasted in engine heat. Car manufacturers plan to have better designs and use lighter weight materials to increase energy efficiency substantially by the year 2000.

Changes Industry Can Make

About 60% of the energy produced at traditional power plants is lost as heat. By using COGENERATION systems, power plants can capture and use the wasted heat to provide steam, hot water, or space heating.

To increase efficiency, some power companies use modifications that cost more per kilowatt-hour to produce but allow consumers to use less energy and pay lower utility bills. As an incentive to get more power companies to make such modifications, regulations were passed in 1989 offering payment to companies for earnings lost due to efficiency modifications. In 1990, the New York Niagara Mohawk Power Corporation ran 12 efficiency programs costing the company $30 million. By saving 133 million kilowatt-hours of power, the

◆ Solar energy is used by some as a means of improving energy efficiency.

Estimated Operating Watts for Common Household Appliances	
electric toothbrush	7
electric shaver	15
hair dryer	600
toaster	1100
coffee maker	600
iron	1100
vacuum cleaner	630
mixer	150
blender	300

To estimate the kilowatt-hour (kwh) for an appliance, multiply the number of watts by the number of hours and divide by 1,000. A kilowatt-hour is a unit of measurement for electric energy.

company was able to get back its $30 million, plus another $1 million in profit.

Many office buildings, covered with tinted glass, are not energy efficient. Windows cannot be opened to let in cool fresh air instead of using air conditioners, and tinted glass prevents solar heat from warming offices in winter. Copy machines, fax machines, electric typewriters, and word processors are often used for small tasks that could be done as efficiently by hand. And some buildings waste energy by leaving on lights at night, when no one is working there.

Newer buildings are being designed to use a minimum of energy. Walls, floors, roofs, and basements are better insulated, and windows are weatherized to reduce heat loss. Timers, room thermostats, and lighting systems that use less energy increase the building's efficiency. For example, the energy efficiency—the amount of visible light—of a fluorescent bulb is 20%, compared to an incandescent, or traditional, lightbulb's 5%.

Changes Consumers Can Make

There are many simple things consumers can do to help conserve energy and make our dwindling resources last longer. Energy-efficient measures used in new buildings can be used in older ones, as well. Homeowners can make sure their houses are well insulated and have adequate storm windows and clean, well-maintained heating and cooling units. Setting heating and cooling thermostats at higher temperatures in summer and lower ones in winter also improves efficiency. And raising window shades in winter allows solar heat to warm a house, if the windows are kept clean; lowering shades in summer keeps out heat and makes air conditioners more efficient. Closing outside doors keeps heat inside in winter and air-conditioned cool air inside in summer. Energy efficiency can be further increased by putting insulated wrap around hot water heaters and pipes so energy is not wasted producing heat where it is not needed.

Turning off unneeded lights, stereos, radios, computers, and TVs improves the energy efficiency of a home. Many people waste energy by opening refrigerator doors and then deciding what they want. Deciding what to get before opening the door, getting it quickly, and closing the door is more energy efficient. Many people use electric appliances, such as toothbrushes, knives, mixers, and dishwashers, when human power does the job just as well. Hanging clothing outside to dry rather than using clothes dryers saves energy, and even drying hair the old-fashioned way, by rubbing it with a towel, saves electricity used for hair dryers. People who cannot do without blow-dryers can be more energy efficient by using them for shorter periods of time.

It is more energy efficient to transport hundreds of people on one train or one bus than to have hundreds of cars transporting one or two people at a time. For shorter trips, people who do not want to ride crowded buses can ride human-powered bikes instead. In addition, because electric motors and batteries are highly efficient energy converters, compared to combustion engines found in most cars, the use of electric cars saves energy. California and the northeast states have laws requiring that 2% of all new cars in 1998 be electric, or "zero-emission," vehicles.

Finally, recycling paper, ALUMINUM, glass, and PLASTICS improves energy efficiency. If every person in a home or business recycles and manufacturers use recycled materials to make new products, less energy is consumed than would be needed to generate new raw materials to produce goods. [*See also* ALTERNATIVE ENERGY SOURCES; CONSERVATION; MASS TRANSIT; and NONRENEWABLE RESOURCES.]

Energy Pyramid

▶ A model that shows the energy relationships between PRODUCERS and CONSUMERS feeding at different TROPHIC LEVELS. All organisms require food and energy to survive. Organisms obtain the energy needed for their survival from the foods they eat. Thus, all organisms within an ECOSYSTEM are related through feeding and energy relationships. These relationships can be shown in a model called an *energy pyramid.*

Each step in the transfer of food and energy from one type of organism to the next in an ecosystem is called a *trophic level.* In all ecosystems, feeding and energy relationships begin with producers, which make up the first trophic level. Producers are organisms that make their own food through either PHOTOSYNTHESIS or chemosynthesis. Common producers are PLANTS, ALGAE, and some types of BACTERIA. Any organism that obtains its food and energy by feeding on other organisms is called a *consumer.* All animals are consumers. Some types of bacteria, protists, and all FUNGI are also consumers.

Because they feed on other organisms, the lowest trophic level at which a consumer can eat is the second trophic level. However, not all consumers feed on the same types of organisms. For example, some consumers feed only on producers. Such consumers are called HERBIVORES. Because they feed at the first consumer level, such consumers are also called *first-order,* or *primary, consumers.* An organism that feeds on other consumers is called a CARNIVORE. When the carnivore feeds directly on an herbivore it is called a *second-order,* or *secondary, consumer.* A second-order consumer feeds on first-order consumers. A consumer that feeds on second-order consumers is called a *third-order,* or *tertiary, consumer.* A consumer that feeds on third-order consumers is a *fourth-order consumer,* and so on.

Some of the energy taken in by organisms feeding at each trophic level is used by the organism to carry out its life processes. Some

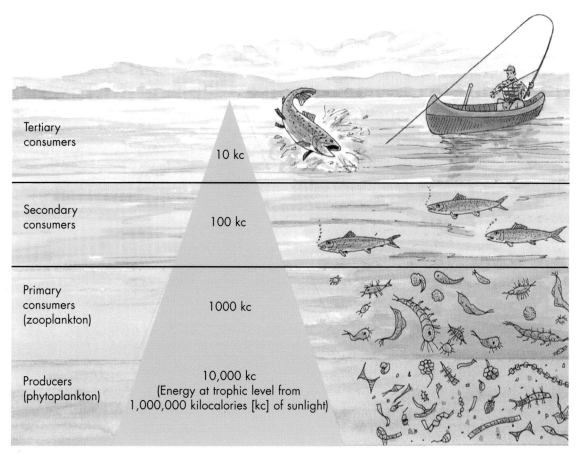

Tertiary
consumers

10 kc

Secondary
consumers

100 kc

Primary
consumers
(zooplankton)

1000 kc

Producers
(phytoplankton)

10,000 kc
(Energy at trophic level from
1,000,000 kilocalories [kc] of sunlight)

◆ Aquatic ecosystems can be analyzed in terms of the energy content of each trophic level. An energy pyramid is a way of presenting the results of such an analysis.

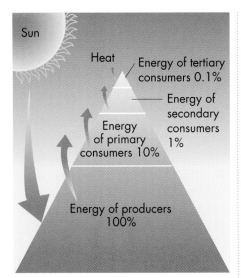

◆ In an energy pyramid, the amount of energy available at each trophic level decreases as you move from the bottom level (the producers) toward the top.

of the energy is also stored in parts of the organism, such as bark, bones, teeth, hooves, and shells, that are not likely to be eaten or digested by other organisms. Still more of the energy is released back into the ENVIRONMENT as heat. Energy used, stored in undigested parts, or given off by the organism is not passed to organisms at the next trophic level. In fact, only about 10% of the energy at any trophic level is passed to organisms feeding at the next level. This loss of available energy is known as the *10% law.*

An energy pyramid illustrates the decrease in available energy at each trophic level. In an ecosystem, 100% of the energy available to the ecosystem is contained in the producer level. However, 90% of this energy is used by the producers, stored in parts of the organism not eaten by other organisms, or given off as heat to the environment.

Thus, only 10% of this energy is passed on to first-order consumers. When first-order consumers are eaten by second-order consumers, only 10% of the energy at the first-order consumer level of the pyramid is passed to the next trophic level. Thus, only 1% (10% of 10%) of the energy contained at the producer level is available to second-order consumers. When third-order consumers feed, they are able to obtain only 10% of the energy from the second-order consumers. Thus, only 0.1% (10% of 1%) of the energy that was available at the producer level is available to third-order consumers. This loss of energy at each trophic level helps to explain why an ecosystem can support more producers than consumers and why the number of consumers at each trophic level is smaller than at the previous level. [*See also* DECOMPOSERS; FOOD CHAIN; and FOOD WEB.]

Environment

◗ A general term that refers to the external conditions in which an organism lives. The word *environment* comes from a French word meaning "to surround." It is used to describe everything that surrounds an organism. The environment includes air, SOIL, CLIMATE, food supply, and a myriad of other external conditions. It even includes the things created by humans.

The word *environment* is sometimes confused with the words *ecology* and *ecosystem.* ECOLOGY is the scientific study of the relationships between organisms and all aspects of their environment. ECOSYSTEM describes the flow of energy through FOOD CHAINS and FOOD WEBS and the other ways in which SPECIES depend on one another in the living world.

◆ The butterfly is part of the rabbit's environment; the rabbit is likewise part of the butterfly's environment.

The Yellow Pine Chipmunk drinks from a river. Water is an essential requirement of all organisms.

The beauty provided by creatures such as these American flamingos is one reason to take care of our environment.

Environmental Education

❱Educational programs designed to increase public awareness of environmental issues. Environmental education seeks to give individuals an understanding of environmental issues and the skills to solve environmental problems.

The Environmental Education Movement in the United States

Environmental education has its beginnings in the conservation education movement of the early twentieth century. At that time, many experts became worried

Students study the different organisms in a pond.

◆ Collecting aquatic specimens is part of a study of lake ecosystems.

about conservation practices on farms and ranches, and about CONSERVATION in general, as the population became concentrated in cities. Educational programs were developed by government agencies and universities to teach people about better practices involving SOIL CONSERVATION, FORESTRY, and other activities involving NATURAL RESOURCES.

After World War II, environmental education expanded. People began to express fears about the impact of new technologies and POPULATION GROWTH on the ENVIRONMENT. During the 1950s and 1960s, concern grew about the impact of nuclear RADIATION from atomic and hydrogen bomb testing sites.

During the 1960s and the 1970s, a variety of efforts and organizations were created with the aim of providing environmental education. Nature centers were started in many communities, and a wide variety of associations were formed at the regional and national levels to help teachers and other educators become better at environmental education.

On April 22, 1970, the first EARTH DAY was held. Earth Day was the brainchild of Wisconsin Senator Gaylord Nelson. Earth Day has been, and still is, an important annual event designed to raise public awareness of environmental issues. Millions of people, businesses, and government agencies become involved in Earth Day activities each year. Environmental awareness increased dramatically after the first Earth Day. The environmental education movement expanded, and many states began to include environmental programs in their public schools.

International Efforts

While environmental education was emerging in the 1970s in the United States, other nations began looking for ways to handle their environmental problems. In 1972, the UNITED NATIONS CONFERENCE ON THE HUMAN ENVIRONMENT was held in Stockholm, Sweden, to discuss international environmental problems. At the conference, it was agreed that international environmental education efforts must be organized and implemented. Three years later, the United Nations approved a $2 million budget to develop a plan for such a program among dozens of nations.

EFFECTIVENESS OF ENVIRONMENTAL EDUCATION PROGRAMS

Today, people become informed about environmental issues through television and radio programs, computer on-line services, magazines, newspapers, and through annual events, such as Earth Day. People also gain a deeper appreciation of nature by visiting the thousands of ZOOS, botanical gardens, aquariums, museums, nature centers, nature reserves, and parks located throughout the world. These public and privately owned institutions offer the public a firsthand look at Earth's BIODIVERSITY, along with information about conservation efforts to protect it.

Environmental education efforts have raised public awareness about global environmental issues. However, critics point out that more education is needed. Often, critics of environmental education state that too much emphasis is placed on environmental problems rather than on possible solutions to the problems. They also point out that most countries still do not have comprehensive environmental education programs to prepare their citizens for the problems of the

◆ A collecting net is used to sample the organisms living in wetlands.

◆ Mountain climbing as part of an environmental education program provides students with firsthand experience in relating to the physical environment.

future. Some critics of environmental education charge that some environmental educators are forcing environmental news on their audiences rather than providing balanced information.

Environmental educators recognize that knowledge, appreciation, and awareness of environmental issues are not enough. Students must also learn the skills and techniques needed to protect the environment. Educators recommend a three-step program for improving environmental education worldwide. The steps of this program are to:

1. provide students with outdoor activities that develop sensitivity toward the environment;

2. show students how to solve environmental problems in their own communities; and

3. reinforce the everyday things that people can do to protect the environment.

A wide variety of environmental education programs now operate in the United States. These range from short programs and presentations to residential camps for school groups or adults. Environmental education can be found in local community nature centers and as part of programs by organizations that offer extended wilderness adventures. A recently formed effort is the National Consortium for Environmental Education and training (NCEET). [*See also* CONSERVATION; ENVIRONMENTAL ETHICS; and UNITED NATIONS ENVIRONMENTAL PROGRAMME (UNEP).]

Environmental Ethics

▌The principles or values people use to decide whether an action that affects the ENVIRONMENT is positive or negative. Many experts believe people should change their ethical standards in order to avoid future environmental crises.

Ethics affect our decisions about the environment. Our moral values determine how we use, treat, and distribute scarce NATURAL RESOURCES. For instance, fresh water is a scarce resource in many parts of the world. Charleston, South Carolina, had a request from tobacco farmers to double the volume of water they withdraw from a river each year. Such a withdrawal would lower the water level in a nearby WILDLIFE refuge. Naturalists estimated this act would halve the number of bird SPECIES nesting in the refuge. Deciding how to reply to the farmers depends partly on ethical beliefs. These beliefs include weighing the relative values placed on growing tobacco, preserving BIRD species, increasing farmers' incomes, and preserving wildlife refuges for recreation.

ETHICS AND ECONOMICS

Values change with time. They also differ from one society to another. In the nineteenth century, the United States had a huge whaling fleet. Now, most Americans dislike the idea of killing and eating WHALES. Norwegians are strong environmentalists on most issues. However, they continue to eat steaks made from minke whales.

Conflicts of values over ethics and economics were at the root of

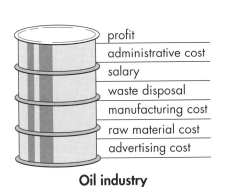

Oil industry

Chemical industry

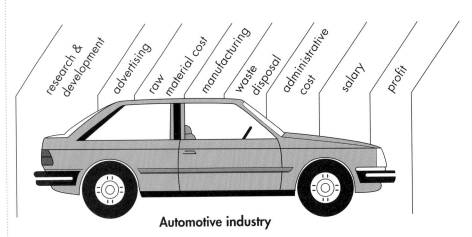

Automotive industry

◆ Corporations view the cost of waste disposal like any other cost. Ethical compromises are made resolving conflicts between incompatible but equally valuable goals—the goal of maximizing profits and the goal of preventing environmental pollution.

◆ Mexico stopped using nets that endanger dolphins in return for compensation from the United States for Mexican fishers.

the argument between the United States and Mexico over the killing of DOLPHINS. Dolphins are not endangered, but many were killed in the nets of TUNA fishing fleets. Money solved this argument and can be used to settle many other environmental disputes. Mexico stopped using nets that endanger dolphins in return for financial compensation to Mexican fishers from the United States.

ETHICAL COMPROMISE

Decisions about the environment often lead to conflicts between incompatible but equally valuable goals. For instance, many people believe that we have a duty to leave a decent environment for future generations. Some people argue that this means we should not feed people who would otherwise die in a FAMINE. If a country has a population that is greater than it can support, people will cut down FORESTS and destroy SOIL trying to grow more crops for food. Unfortunately, these acts increase the chance that more people will starve to death in the future. The opposing argument is that it is unethical to permit people to starve to death if we have the means to save them.

The practical solution to such ethical conflicts usually requires some sort of compromise. In the North African famines of recent years, some of the aid supplied by concerned people all over the world has been used to feed the starving. However, some aid is being used for education and projects designed to prevent famine in the future. Such projects focus on

agricultural advice, reforestation, and FAMILY PLANNING. In this particular case, many observers believe that too little of the aid has been spent on population control. These people argue that so little fertile soil and fresh water exist in north Africa that famines and starvation will continue unless the population is reduced.

One of the leading early thinkers on the subject of environmental ethics was Aldo LEOPOLD, a wildlife researcher and professor who published a variety of articles and books about our relationship to the environment in the mid-1900s. In his book entitled *Sand County Almanac*, he argued that people must extend similar ethical consideration to the environment that they do to other people. He called this idea the "land ethic." [*See also* FRONTIER ETHIC; OVERPOPULATION; and SUSTAINABLE DEVELOPMENT.]

Environmental Impact Statement

▌▶The evaluation of potential environmental effects conducted under the NATIONAL ENVIRONMENTAL POLICY ACT (NEPA). Signed into law in 1970 by President Richard M. Nixon, NEPA mandated the evaluation of environmental impacts of major federal actions. Under NEPA, the COUNCIL ON ENVIRONMENTAL QUALITY (CEQ) was created to develop guidelines and regulations for environmental impact assessment.

Under NEPA, major federal actions were defined as activities over which the U.S. government had control, either because it provided funds or because it gave permission for certain actions to be taken. Federal agencies were directed to develop their own rules and guidelines for their actions that required an environmental impact statement. For example, the Department of Defense has rules that require an environmental impact assessment to be made if a military base is opened or moved.

NEPA requires federal agencies to evaluate the environmental impact of their activities in a specific way. Agencies must provide alternative ways of accomplishing a project. Under NEPA, agencies must always evaluate the environmental impact of taking no action. Assessing the environmental impact of a project includes evaluations of all impacts on the "human environment."

The procedure for producing an environmental impact statement (EIS) is specified by the CEQ's rules. In general, any agency must take the following steps:

1. Notice of intent. Agencies place a notice of intent in the Federal Register stating their reasons for proposing a particular project.

2. Scoping. Once the notice of intent is published, the agency begins a process called *scoping*, in which the important issues to be evaluated in the EIS are determined. Often, public hearings or information meetings are held, and letters are received from individuals, public interest groups, and other government agencies. These activities determine the scope of impacts to be evaluated. That is, important concerns are identified, and issues considered to be unimportant are likely to be ignored in the impact assessment process.

3. A Draft Environmental Impact Statement (DEIS). After the scope of the impact statement is determined, the DEIS is written. The contents of the DEIS are expected to include a description of the alternatives to the project that are being considered (including taking no action), the ENVIRONMENT that might be affected, and the reasons for selecting the "preferred alternative." The preferred alternative is the option the agency is likely to follow. Each alternative is supposed to be evaluated for its potential impact on the environment. An environmental impact may be the effects on ECOSYSTEMS, the local economy, historical buildings, or other parts of the "human environment."

4. Public comment. The DEIS must be circulated to citizens and agencies that wish to comment on the action the agency is planning. Public comment comes in written form and from one or more public meetings, usually at a place near the project site.

5. Final Environmental Impact Statement. Once comments are received, the agency has an opportunity to revise the DEIS before issuing its final evaluation. The final

◆ The repercussions of building a hydroelectric dam are discussed in an Environmental Impact Statement required by the law.

evaluation must also contain responses to the comments received in the public comment period.

6. Record of decision. Once the agency decides what, if any, action it will take on a project, it issues a record of decision, and the impact assessment process ends.

There have been many criticisms of NEPA. One of the criticisms is that NEPA requires only that the federal government evaluate environmental impacts of its own actions and does not provide the same protections for state, local, or private projects. In addition, critics complain that the NEPA contains no enforcement methods. Citizen groups and other government agencies have sometimes had to file lawsuits against agencies for failure to follow NEPA rules and regulations. Because lawsuits often take many years to be settled, the fear of being sued is one of many reasons federal agencies usually follow NEPA rules.

Some agencies have modified their own rules for complying with NEPA. For example, the Department of Transportation gives money to states for building roads, and each road project requires an environmental impact assessment. Every possible alignment of a new road could be considered an alternative to be evaluated. To save time, transportation agencies often have a large number of road alignment options available for comment during the scoping period to find out which alternatives may seem to be the most controversial. In this case, the scoping process also helps remove some alternatives from consideration.

◆ When planning to build a major highway, a state is required to submit an EIS before it can receive federal funding.

Some agencies have developed lists of things they do routinely and have excluded these activities from the impact assessment process. For example, the NATIONAL PARK SERVICE does not evaluate the effects of routine maintenance on parking lots or roads, even though these expensive activities might be considered major federal actions.

When the importance of a project is not clear, an agency can conduct an environmental assessment, a short form of the EIS process. For example, in 1988 the National Park Service began to develop the site of the lake that produced the famous flood in Johnstown, Pennsylvania. Part of the old lake bed had been planted previously with pine trees. The National Park Service conducted an environmental assessment to determine the potential

effects of removing the trees. If significant impacts were predicted, an EIS would need to be produced. If the impacts were small, the plan to remove the trees would not need a major assessment. When the National Park Service finished the environmental assessment, it issued a Finding of No Significant Impact (FONSI) and proceeded with the project. The FONSI stated clearly that no adverse impacts were expected and that an EIS would not be issued.

An EIS is not a scientific document, and sometimes this fact has led to confusion. An EIS is a legal document indicating the assessment of impacts. Because an EIS is written in nontechnical language, scientists and other experts sometimes criticize it for its lack of specificity. An agency is not required to choose

have developed other processes, such as issuing permits, to avoid the long and expensive impact assessment process. For example, when a city builds a new SEWAGE TREATMENT PLANT with money that comes from the ENVIRONMENTAL PROTECTION AGENCY (EPA), the environmental effects are usually restricted to setting limits on permits to discharge the treated WASTEWATER, rather than to preparing an environmental impact statement.

NEPA and the environmental impact statement process have had strong positive effects on controlling environmental impacts. Many projects are evaluated for adverse effects using the NEPA model of comparing the impacts of several project alternatives before a decision is made. Often, this process involves the general public, as well as specialists.

◆ An environmental impact statement must be developed for siting and constructing a new nuclear power plant.

Environmental Justice

▌A movement seeking to minimize the effects of environmental POLLUTION on particular disadvantaged human groups. An adverse environmental impact can affect some groups of people more than others. For decades, people have come to larger towns and cities seeking opportunities. Often, these people can afford to live only in areas with low rents and, sometimes, substandard housing. These areas of cities are usually those with

the alternative that produces the least environmental effect; only the assessment of effects is required. Agencies often pay much more attention to assessing the effects (good and bad) of their preferred alternative at the expense of other options. Still, NEPA has assured that the impacts of major federal projects like DAMS, NUCLEAR POWER plants, highways, and other projects are

evaluated. Many states have adopted similar laws for state projects.

The number of EISs has been dropping. The first EISs were very large and expensive projects, often because the local environment was exhaustively described. In 1976, the CEQ adopted new rules limiting the size of EISs. Over the years, many agencies have found other ways to evaluate and control impacts and

◆ Illegal dumps are often established in low-rent neighborhoods. The owners of this property in Bridgeport, Connecticut, have been fined $900,000 for this illegal dump and have been told to remove it.

major industries or are heavily affected by transportation, such as railroad switching yards or commercial airports. These areas may also have frequent spills of hazardous materials and may have accumulated the wastes of generations, resulting in the potential exposure to hazards for people who must live there.

The realization that high-density industrial development or urban decay often goes hand in hand with settlement by poor people has suggested that some groups in society bear an unfair share of exposure to pollutants. Activists in the environmental justice movement point out that this unfair share of pollutant exposure affects poor people in general, and certain racial and eth-

nic minority groups in particular, who may be restricted to the dirtiest areas of cities.

One of the most common examples of an environmental justice problem is the exposure of inner city children to LEAD from lead-based paint. Small children may eat flakes of lead-based paint and many children and adults may breathe in dust with high lead concentrations in old areas of cities. Lead poisoning is known to cause a variety of health and development problems, especially learning deficiencies. Lead paint has rarely been used in houses built or remodeled since the 1970s. Therefore, potential lead poisoning is seen as an environmental justice issue because families living in the suburbs (pre-

dominately white) are not exposed to the same hazards as poor families (primarily African Americans, Hispanics, and immigrants) in the inner cities. Parallels can be drawn for exposures to potentially hazardous materials used in industry.

On a national level, the ENVIRONMENTAL PROTECTION AGENCY (EPA) has played an important role in environmental justice studies and mitigation efforts. On a local level, many towns and cities have land-use planning and zoning restrictions to limit the contact between living areas, commercial areas, and industrial areas. However, zoning is not a perfect solution, since the people forced to live in low-cost housing are often on the border between housing areas and polluted areas. Advocates for environmental justice are concerned about locating potential adverse impacts and tracking human populations negatively affected by pollution. Similarly, workers earning low wages may be exposed to hazards not experienced by better-paid workers. The study of environmental justice attempts to locate and solve problems of human exposure to environmental pollutants by revealing the level of exposure and working to minimize or eliminate effects on minority and low-income people.

Environmental Protection Agency (EPA)

▌The agency of the United States government that is responsible for enforcing many federal environmental laws, particularly those concerned with impacts on human health. The U.S. Environmental Protection Agency (EPA) was established by President Richard Nixon in 1970 as part of the executive branch of the U.S. government. Prior to 1970, the federal enforcement of environmental laws was split among many parts of the executive branch, including the U.S. DEPARTMENT OF AGRICULTURE, the U.S. ATOMIC ENERGY COMMISSION, and the U.S. DEPARTMENT OF THE INTERIOR. The EPA enforces most federal environmental laws. However, it does not regulate concerns about NATIONAL PARKS, NATIONAL FORESTS, wilderness areas, WILDLIFE refuges, and some other PUBLIC LANDS.

ENFORCEMENT

The executive branch of the government is headed by the President. Its role is to enforce laws. The EPA enforces laws by doing several things. First, it must write regulations, or rules, that help define the law. For example, regulations may spell out how much POLLUTION an area may have and how a polluting facility, such as a factory, can get POLLUTION PERMITS.

Issuing Permits

The EPA issues pollution permits to companies. These permits allow a company to give off a specified amount of pollution to the air, water, or on land. After a permit is issued, the EPA makes periodic checks to make sure a facility does not give off more pollution to the ENVIRONMENT than its permit specifies. It takes many people to carry out these checks because the

◆ The EPA monitors emissions from industrial sites such as this phosphate fertilizer plant.

United States is so industrialized. Congress provides the monies needed for the salaries of inspectors. It may also pay fees to attorneys hired to file a case against a company that does not live up to the conditions of its permit.

Enforcement of the Clean Air Act

The CLEAN AIR ACT OF 1990 is one law the EPA enforces. This act requires the EPA to establish National Ambient Air Quality Standards for seven outdoor air pollutants. These pollutants include suspended PARTICULATES, sulfur oxides, CARBON DIOXIDE, nitrogen oxides, OZONE, HYDROCARBONS, and LEAD. Ambient air is the air in your surroundings. Two kinds of standards were established: primary, which are to protect human health; and secondary, which are to protect crops, buildings, water supplies, and to maintain visibility in the air.

Enforcement of Acts Dealing with Water

The SAFE DRINKING WATER ACT OF 1974 requires the EPA to establish standards of pollutants in drinking water. These standards are expressed as "maximum contaminant levels." Two laws that control the pollution of SURFACE WATERS— the CLEAN WATER ACT and the Water Quality Act—mandate that the EPA establish national effluent standards.

An effluent standard is a measurement of how much pollution may be in any substance that flows from a factory. The EPA issues permits to companies stating how much pollution each may release into waterways. The law also applies to the effluents from SEWAGE TREATMENT PLANTS. In 1989, the EPA found that 66% of the nation's sewage treatment plants exceeded the standards.

Enforcement of Laws Controlling Toxic Substances

The FEDERAL INSECTICIDE, FUNGICIDE, AND RODENTICIDE ACT OF 1972 is enforced by the EPA. To enforce this law, the EPA needed to evaluate more than 600 active ingredients in PESTICIDES. The EPA did not have the funds needed to carry out this task, so the agency made use of information regarding pesticide ingredients provided by pesticide manufacturers. The EPA originally used information provided by the pesticide manufacturers to determine possible hazards, since insufficient funds were available to test the large and growing number of pesticides. In the late 1980s, the EPA began the long process of reevaluating the dangers posed by pesticides. This is a long and ongoing task.

The TOXIC SUBSTANCES CONTROL ACT controls the activities of chemical companies. To make a new chemical, a company must get a permit from the EPA. The EPA studies data about the chemical. It may then issue a permit or ban the substance if it is found to threaten the environment.

THE SUPERFUND

The EPA administers the law called the SUPERFUND. This law is also known as the COMPREHENSIVE ENVIRONMENTAL RESPONSE, COMPENSATION, AND LIABILITY ACT OF 1980 (CERCLA). Under CERCLA, monies from taxes on chemicals are used to identify TOXIC WASTE disposal areas that threaten to pollute groundwater. Additional monies are then set aside and used to clean up these areas. The Superfund process is long and involved. The EPA is often criticized for listing and cleaning up too few sites.

THE ROLE OF THE EPA IN THE NATIONAL ENVIRONMENTAL POLICY ACT

The EPA plays an important role in the NATIONAL ENVIRONMENTAL POLICY ACT (NEPA). As part of this act, the EPA participates in the ENVIRONMENTAL IMPACT STATEMENT (EIS) process. The EPA reviews and comments on EISs written by any federal agency concerning almost all environmental actions that may have impact on WATER POLLUTION, AIR POLLUTION, drinking water supplies, SOLID WASTE, pesticides, RADIATION, and noise. It frequently reviews EISs of state and local government offices.

A list of EISs and the EPA comments are published in the Federal Register so all citizens of the United States may become aware of projects affecting the environment. Based on information it receives in an EIS, the EPA either endorses a project, asks for more information, or objects to the project. If it objects, the objection is published and the matter is referred to the COUNCIL ON ENVIRONMENTAL QUALITY. The EPA does not have the authority to stop a project. It acts only in an advisory capacity.

Erosion

❚❚The wearing away of Earth's land by natural forces such as wind and water. Erosion shapes Earth and moves materials on its surface from one place to another. It can be very destructive. For example, it can remove the rich TOPSOIL needed to grow crops. Beach erosion is another significant problem. Such erosion destroys beachfront that is valued for its beauty.

Earth's surface is composed of rock and SOIL. The surface is changed as rock crumbles to form rock particles, a major ingredient of soil. This breaking down and wearing away of rock by water, chemical abrasions, and the action of PLANTS and animals is called WEATHERING. Soil formed from rock particles is then moved about by the erosive forces of moving water, wind, glaciers, **gravity**, and ocean waves.

EROSIVE FORCES

Rainwater, pelting down on soil, moves down slopes, carrying soil away in streams of water. If there is nothing to slow the water or hold the soil, flowing water can begin to erode sharp-banded gullies. On a larger scale, erosion by water creates and shapes stream and river channels.

The amount of soil particles, called SEDIMENT, carried by a river each year is measured in tons. For example, at its mouth, the Mississippi River carries away 344 million tons (310 million metric tons) per year. The Amazon in South Amer-

◆ The removal of vegetation from mountainsides results in loss of topsoil because there are no roots to hold the soil in place against the eroding power of wind and water.

◆ Plants growing on an eroding dune will soon die as the soil is removed.

ica moves about 500 million tons (450 million metric tons) of soil each year. Under natural conditions, about 1 to 5 tons of soil per acre is eroded per year.

If unusually large amounts of sediment are carried into a river, it can gradually fill up RESERVOIRS and lakes, clog the gills of FISH, smother fish eggs and small hatchlings on the stream or lake bottom, and significantly alter the HABITAT of some aquatic organisms.

Unlike water, wind can move soil both uphill and downhill. In a secondary erosive effect, the particles moved by the wind are abrasive. They erode soil as they move and wear down rocks and loose stones. Like water, wind is capable of moving fine soil particles over great distances. Powerful desert winds have moved soil from North Africa hundreds of miles north to the decks of ships in the Mediterranean and even to areas in southern Europe. During the 1930s, dust storms in the western United States carried soil from Kansas and Oklahoma to the East Coast of the United States.

Glacial erosion, or GLACIATION, takes place mostly in cold, high mountain ranges. GLACIERS, moving masses of ice and snow, formed during the last ICE AGE half a million years ago. During this period, ice covered Europe from the North Pole to as far south as central Germany and Great Britain.

In North America, all of Canada and the northern United States was covered with ice. Today, glaciers move at speeds varying from a few inches to many feet per day, creating and shaping mountain valleys as they move. Soil and rock may be deposited at times to the side and front of the glacier. Large blocks of rock carried in the lower part of the glacier cut grooves in the rocky floors of mountain valleys. Meltwater from the glacier is made milky by the rock particles that are ground away by the glacier's movement.

Gravity is also an erosive force. Gravity pulls rock and soil downhill. Movement caused by gravity may be gradual. Such movement is called a *creep* or a *landslide.* The tilting of trees, poles, and gravestones on a hill is an indication of creep. Landslides usually occur in mountainous areas such as the Swiss Alps and the Asian Himalayas. A landslide may move soil in amounts ranging from a few cubic yards to several cubic miles. The landslide may block a stream, creating a new lake, or cover a small town or highways.

In flat areas in Norway and Canada, a kind of soil called *quick clay* can move quickly from one area to another. This happens because rain and groundwater wash away the salt that binds the soil particles together. Without warning, the clay begins to move quickly down even the smallest slope. In 1893, in south-central Norway, a quick clay slide of 72 million cubic yards (55 million cubic meters) covered a 3.3-square-mile (8.5-square-kilometer) farming area within about half an hour. Twenty-two farms were destroyed, and 111 people were killed.

Ocean waves, pushing in on coasts and then retreating, move sand. Waves usually come into a beach at an angle, picking up sand and carrying it in the direction they are moving. The direction in which waves move changes with the seasons. This varies the distribution of the sand over time. High waves, produced by a storm, pull sand off a beach and deposit it on the ocean bottom. A single large storm can tear away as much as 50 feet (15 meters) of sand from the width of a beach. Gentle waves usually move sand onto the beach. The back and

◆ Without vegetation holding soil in place, it is washed down a country road in heavy rain.

forth action of storm waves and gentle waves can widen or narrow the beach by as much as 100 feet (30 meters).

Storm surges, which are created when strong winds blow across water, push the water far onto the shore. Waves from storm surges can also carry away large quantities of inland sand and dunes.

RESULTS OF EROSION

Erosion has many effects on Earth's organisms. It is a huge problem for farmers when valuable topsoil is carried away. Topsoil provides the nutrients needed to grow crops. Large areas of the farmland east of the Mississippi River in the United States lose between 5 and 13 tons (5 to 13 metric tons) of topsoil per half acre every year. In the west, the toll is lower—2 to 5 tons per acre (2 to 5.5 metric tons per half hectare). In contrast, topsoil forms very slowly. It is the product of years of interaction with rock dust, rain, sun, and decayed plant and animal matter. Agriculturists estimate that it takes 100 to 1,000 years to make only 1 inch (2.5 centimeters) of topsoil.

While erosion is a natural process, many farming methods can make it much worse. For example, plowing turns a compact soil surface into loose particles. Cultivated plants use up the organic matter in the soil—organic matter is the glue that holds the soil particles together. Neat rows of crops, unlike natural dense plant cover, leave much of the soil bare and vulnerable to erosion.

Many farmers have adapted their farming methods to prevent or at least reduce erosion. For example, soil around plants is often covered with a protective layer of

◆ Rainwater rushing down small depressions in a field can eventually produce bizarre gullies.

◆ The next powerful storm may further erode the groundfloor under the houses and trees in Isle Vista, California, causing them to be swept away into the ocean.

shredded plants and plant parts called *mulch*. Mulch helps hold loose soil in place and returns nutrients to soil through DECOMPOSITION. Grassy strips that stop RUNOFF are planted at intervals in fields of row crops. Small earthen DAMS are also used to stop water flow. CONTOUR FARMING produces rows that circle a slope. Ridges left by the plow act as small dams, catching water as it moves downhill. On steeper slopes, crops are planted in rows with such crops as corn alternated with cover crops such as grain. The thick grain crop catches

the soil and water as they move down the slope.

Coastal dwellers face serious problems when erosion carries away large parts of a beach. The net effect of wave action is the eroding or narrowing of beaches. The United States has about 36,000 miles (58,000 kilometers) of coastline. Nearly half of this is eroding. Beach erosion is worse on the stormy Atlantic coast. In a 30-mile (48-kilometer) length of coastline of western England, wave action has cut the coastline 3 miles (4.8 kilometers) inland in the past 1,300 years. The sites of a number of villages have simply disappeared beneath the OCEAN. Ocean waves, carrying rocks and rock fragments, also erode high cliff shorelines, undercutting them at the bottom. Eventually, the top of the cliff may topple into the ocean.

A number of fairly successful methods of preventing beach erosion are now being used. Concrete seawalls are built to protect the beach from wave action. Large rocky **breakwaters** are placed in the water parallel to the beach. Jetties are built perpendicular to the beach to trap sand moved by waves. Sand fences and salt-tolerant plants, such as **sea oats** and beach grasses, are planted to hold the sand dunes behind the beach in place.

A beach can be restored by trucking in large amounts of sand dredged from offshore. This process reverses the work of the waves. Producing a wide beach is the best protection against the effects of beach erosion. [*See also* AIR POLLUTION; CLEAR-CUTTING; DUST BOWL; GULLYING; NO-TILL AGRICULTURE; SOIL CONSERVATION; TERRACING; and WATER POLLUTION.]

Estuary

▶An area, such as the mouth of a river or an inlet, where fresh water and salt water mix. Fresh water is less dense than salt water. At the mouth of a river, fresh water flows from the river into an OCEAN. At the same time, a wedge of salt water from the ocean moves in toward the land. This dense salt water flows under the less dense fresh water. The place where these two types of water meet and mix forms an estuary.

ESTUARY FORMATION

Currents are set up at the place where fresh water and salt water meet. The currents cause SEDIMENT carried in the moving water to fall and slowly fill the estuary. On the East Coast of the United States, estuaries average about 12 feet (3.67 meters) in depth. A deeper channel forms in the middle of the estuary, where mud is washed away by the flow of the river.

Many estuaries, such as the Chesapeake Bay, are actually drowned river valleys. A valley is drowned when it partly fills with water because the sea level rises or the land subsides. Estuaries also form as coastal sand dunes are washed away from the land by ocean waves. If the removal of sand causes the coastline to sink, an estuary forms. At the same time, sand dunes are formed in the ocean by the dropping of sand carried in ocean waves. These sand dunes form BARRIER ISLANDS. Barrier islands are islands that form off a coast, like

those along most of the eastern and gulf coasts of the United States.

Barrier islands are battered on the ocean side by wind and waves. However, on the estuary side, water moves slowly and is shallow. The traits of barrier islands allow them

◆ Salt marshes form where rivers deposit their load of mud in estuaries. They serve as a nursery for young animals, providing food and protection.

◆ The Salmon River estuary of Oregon is a place where fresh water from the river comes in contact with salt water from the Pacific Ocean.

to develop distinctive vegetation. The PLANTS have deep roots that can withstand the carrying away of sediment by currents and have tough stems that are able to tolerate salty wind and water on the seaward side and marsh on the other side.

IMPORTANCE OF ESTUARIES

Estuaries provide protected harbors, access to channels to the ocean, and connections with river systems. Not surprisingly, they are the sites of many of the world's major ports. Of the ten largest urban areas in the world, seven border on estuaries. These include Tokyo, New York, Shanghai, Calcutta, Buenos Aires, Rio de Janeiro, and Bombay.

Because of their beauty, estuaries often become desirable places for people to live. In some areas, people have built houses on the seaward sides of barrier islands. Such construction is common in states like Florida and New Jersey. These areas provide people living there with beautiful views of the ocean and recreational activities, such as swimming, fishing, and boating. However, because of their nearness to the ocean, such areas are vulnerable to storms. Building in such areas destroys sand dunes that normally protect sandy shores from EROSION.

ESTUARY ECOSYSTEMS

Estuaries are productive ECOSYSTEMS because they constantly receive fresh nutrients from both the river and the ocean. In addition, because they are partly surrounded by land, estuaries are protected from the

◆ Estuary creeks and their salt marshes are found worldwide. This is an aerial photo of Fraser Island, Queensland, Australia.

worst force of ocean waves. In parts of the estuary where water is shallow enough, plants may root and begin to grow in the mineral-rich sediment deposited by ocean and river waters. Over time, a WETLAND, or a land area that is flooded all or part of the time, forms.

Eastern SALT MARSHES form where rivers such as the Mississippi deposit their load of mud and nutrients in estuaries. Smooth cordgrass covers thousands of acres of the marsh, with scattered strands of other plants on higher, drier ground. The marsh serves as a nursery that provides food and protection to many young animals. For instance, commercial shrimp in the southern part of the United States live in the oceans as adults and lay their eggs on the ocean bottom. These eggs are carried toward the shore by tidal currents until they reach the marsh, where they hatch. The young remain in the marsh until they grow to adulthood. Then the adults once again return to the ocean. Other animals that spend part of their lives in salt marshes include the young of several SPECIES of crabs, as well as many species of FISH. As these animals grow to maturity, they migrate down river and eventually out to sea. In the open ocean, these animals become part of the ocean's FOOD CHAIN, where they may become food for larger fish such as swordfish, snapper, grouper, and TUNA.

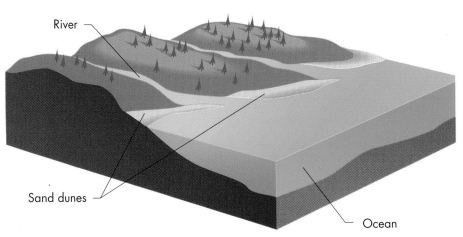

Sea level rises:

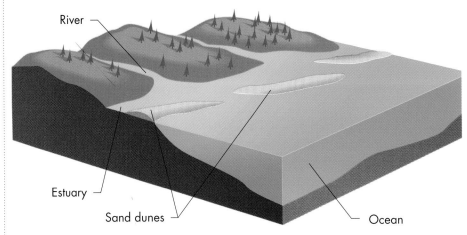

◆ Many estuaries are drowned river valleys that were drowned by rising sea levels or by subsiding land as illustrated above.

ESTUARY DAMAGE

About half of all the coastal wetlands and estuaries in the United States have been destroyed by the actions of people. The situation is even worse in many other countries. In the Philippines, MANGROVE swamps support a huge harvest of fish and shellfish. However, plentiful harvests are decreasing as the swamps are destroyed. In Puerto Rico, MINING for sand and construction of an airport have destroyed large areas of mangrove swamp.

Along the coast of the United States, LANDFILL operations, together with DREDGING for sand and gravel, have destroyed 12,428 square miles (20,000 square kilometers) of wetlands. In addition, wetlands are often viewed as convenient places to dump wastes. This practice leads to POLLUTION of both the land and the water. Currently, about one-third of the estuarine regions in the United States are closed to shellfish collecting because the regions are polluted. Once a wetland has become polluted and begins to fill in, it is often used as a building site.

The pollutants that damage estuaries and wetlands are the same as those that pollute other waterways: SEWAGE, TOXIC WASTE, and agricultural RUNOFF of soil, PESTICIDES, and fertilizer. Aquatic ecosystems can degrade all of these pollutants, with the exception of some toxic wastes. However, these ecosystems cannot degrade materials in the quantities produced by dense human populations. Off the coasts of some parts of New York and New Jersey, for instance, much sewage is dumped each year. These wastes destroy the ecosystem of the continental shelf. The native species of these areas are then replaced by organisms that can live in high concentrations of sewage. In the past, New York and New Jersey dumped much of their GARBAGE at sea. This practice drew protests in 1988 and 1989 when MEDICAL WASTE, including syringes and bags that once contained blood, started washing up on New Jersey beaches.

Most coastal wetlands form in estuaries. Coastal wetlands, such as salt marshes and mangrove swamps, are needed for the survival of valuable seafood species. In addition, these wetlands absorb large amounts of pollution. An example is the salt marsh that covers much of the shoreline of the Gulf of Mexico and the Atlantic coast of the United States. Salt marsh also once made up large areas of the California coast. However, this salt marsh was almost completely destroyed by development. California now has plans to restore some of its estuarine wetlands.

In recent years, some communities that are built around estuaries have worked to protect their beaches from erosion. Others have prohibited construction near the ocean, restored damaged dunes, and built walkways over dunes to prevent vegetation that holds the dune in place from being destroyed by people walking on it. Such efforts and others are being made to help restore estuary regions to their original condition.

Eutrophication

▶A process in which a pond or lake becomes thick with vegetation and other organisms due to high levels of nutrient in the water. Natural eutrophication takes place at rates varying from decades to thousands of years, but human activities such as SEWAGE disposal and land drainage, can speed up the process. This has had a major effect on ECOSYSTEMS in thousands of lakes and waterways throughout the industrialized regions of the world.

When PHOSPHATES from DETERGENTS or fertilizers containing nitrogen enter rivers, streams, and lakes, the sudden increase in the nutrient supply creates excessive algal growth, or ALGAL BLOOM. When the ALGAE die, bacterial DECOMPOSITION may lower the oxygen level of the water. Low oxygen levels may cause the deaths of FISH and other aquatic animal life. As these animals decompose, there is further depletion or even elimination of OXYGEN from the water. A eutrophic pond or lake can have very low DISSOLVED OXYGEN, especially in winter when a great deal of decomposition occurs. A body of water that quickly becomes eutrophic can have large parts of its fish population die at this point. Eutrophication severely affects the ability of the pond or lake to support some kinds of organisms. [*See also* BIOCHEMICAL OXYGEN DEMAND (BOD); DISSOLVED OXYGEN; EFFLUENT; NITROGEN CYCLE; SEWAGE; and WATER POLLUTION.]

◆ Eutrophication occurs when an oversupply of nutrients enters a body of water, causing algae to grow excessively. The nutrients may be from such sources as sewage disposal or land drainage.

Evapotranspiration

▶ The total process of transfering water from vegetated land into the ATMOSPHERE. Evaporation refers to the change of liquid water into water vapor. Evaporation occurs regularly from all bodies of water, such as lakes, streams, rivers, and OCEANS. It also occurs from wet surfaces such as wet SOIL. Transpiration is the process in which water is absorbed from the soil by a plant's root system, transported up through the plant's vascular system, and evaporated off a plant's surfaces, particularly its leaves.

The processes of evaporation and transpiration play an important role in the global WATER CYCLE. For example, some tropical trees may evapotranspirate more than 100 gallons (378 liters) of moisture per day into the atmosphere. Together, they help return water from the land to the atmosphere, where it will be redistributed to the land once again as rain, snow, and other types of PRECIPITATION. [*See also* BIOGEOCHEMICAL CYCLE and PLANTS.]

◆ Evapotranspiration occurs in all plants having microscopic tubes, like streams, that carry water up from roots to leaves. Red clover has such tubes.

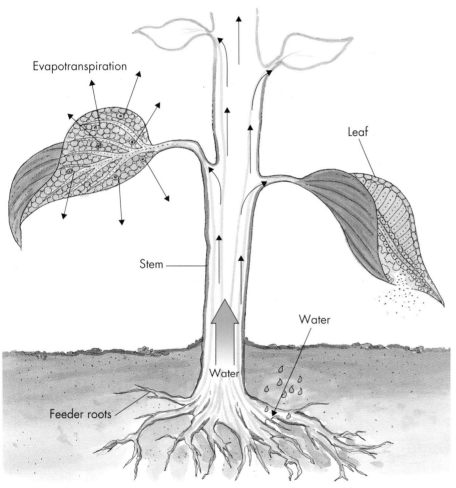

◆ Water and minerals move upward through microscopic tables in plants as a result of evapotranspiration, the evaporation of water through leaves. Water molecules form an unbroken chain due to their attraction to one another. Evapotranspiration helps to pull the entire chain upward and bring new water molecules into the roots.

Everglades National Park

▶ A NATIONAL PARK located in South Florida. The park makes up a portion (1.4 million acres, 566,000 hectares) of the WETLAND region called the Everglades, a region that extends from Lake Okeechobee to the southern tip of the state.

The Everglades is primarily composed of a very slow river, averaging 6 inches (15 centimeters) deep and 50 miles (80 kilometers) wide, that flows south to the OCEAN. Because of the great expanses of vegetation that grow in that river, the Everglades is also referred to

as the "river of grass." The Everglades is a unique ECOSYSTEM due to the mix of temperate and subtropical SPECIES found there and the great number of animals that make it their home. It was the first national park in the United States to be set aside for its biological richness rather than for its geologic features.

The Everglades is not a single ecosystem. Some parts of the Everglades are freshwater wetlands. Near the southern coast, MANGROVE swamps thrive in slightly salty water and support distinct and abundant plant and animal life. Water flows into the Everglades from Lake Okeechobee. The water flows slowly over the whole region, and its depth may vary from a few inches to about 6.5 feet (2 meters). The rainwater that spills out of Lake Okeechobee is seasonal. The wet season is from May through October, when the region is (or should be) inundated by water. Fall and winter are dry months. During this period, many areas of the Everglades dry out, allowing natural fires to burn and regenerate grasses.

THREATS TO THE EVERGLADES

Florida's population has been growing as retirees, sun lovers, and immigrants settle in the area to enjoy the CLIMATE. In fact, Florida has one of the highest rates of POPULATION GROWTH in the nation. Industries, too, have been moving there. All this population growth has had disastrous effects on the Everglades. To save the only parcel left of this irreplaceable natural treasure, the Everglades National Park was established in 1947. However, the degradation of the Everglades continues today.

To create farmland, water was channeled from Lake Okeechobee. A maze of canals was built to direct the lake's water to nearby farms. Water conservation areas were established around Lake Okeechobee to provide water to farms. As a result of these practices, the Everglades suffered. Vast amounts of wetlands were destroyed because the water supply was cut off.

◆ The map shows the Florida Everglades.

In 1967, a canal was dug from the lake to the national park area of the Everglades in an effort to bring in more water. However, the canal could not deliver the flow of water the area needed because following a rainfall, water rushes down the narrow canal and floods much of the Everglades. Supplies are also limited because the park receives water only when officials decide to release water from the conservation areas. Thus, the Everglades suffers a periodic shortage of water that is held back in the conservation areas, as well as periods of destructive flooding when water is released.

EVERGLADES FLORA AND FAUNA

Hundreds of species of PLANTS and animals once abundant in the Everglades are now extinct. More than 400 of the plant and animal species that survive are endangered. Some biologists estimate that at least 50% of the birds that once nested in the Everglades are gone. For some highly ENDANGERED SPECIES, breeding populations have declined more than 90%.

AGRICULTURAL EFFECTS

Agriculture is damaging the Everglades in another way, too. Lake Okeechobee is surrounded by sugarcane farms and cattle and dairy ranches. Fertilizer from the farms and animal waste from the ranches wash into Lake Okeechobee. Both the fertilizer and animal waste contain nutrients that promote plant growth. However, the native plants necessary for animals to survive are being crowded out by EXOTIC SPECIES such as European cattails. The excessive plant growth is also changing what were once very "wet" wetlands into drier, upland areas that are alien to the native Everglades ecosystem.

SAVING THE EVERGLADES

The U.S. government is suing the state of Florida to get the state to enforce laws designed to preserve this ecosystem. In 1983, Floridians and officials in Everglades National Park launched a "Save our Everglades" effort. Since then, more than 100,000 acres (40,000 hectares) of the Everglades have been restored to wetland. Preservation of the Everglades is of keen interest to many people. Besides being a national park, it has also been declared a Biosphere Preserve by the United Nations, recognizing its importance to global BIODIVERSITY. [*See also* AQUIFER and NATIONAL PARKS.]

◆ Many rivers run through the Everglades.

THE THEORY OF EVOLUTION

The theory of evolution accepted by most scientists was developed by Charles DARWIN in 1859. Darwin published his ideas in his book, *On the Origin of Species*. According to Darwin's theory, the evolution of species involves four main factors. These factors are overproduction; COMPETITION; variation; and NATURAL SELECTION, or survival of the fittest.

Overproduction

Simply stated, overproduction refers to the tendency of organisms to produce many more offspring than can survive and reproduce. This tendency is well illustrated by egg-laying FISH. Such fish often release millions of eggs into the water during mating season. However, not all of these eggs are fertilized by a male fish of the same species. Those that are not fertilized do not develop. Of the eggs that are fertilized, many are eaten by other organisms. A smaller number of eggs develop and hatch into young fish. Of these hatchlings, many are eaten or die from starvation or the effects of pollutants or toxic chemicals in the water. An even smaller number will survive to adulthood and reproduce.

Competition

The NATURAL RESOURCES in any given ECOSYSTEM are limited. Thus, organisms must compete with each other for such resources as food, living space, and water. For example, PLANTS living on the forest floor must compete with tall trees for the sunlight needed to carry out

◆ The Everglades National Park is located in the southernmost part of the Everglades.

Evolution

▮Process of change in the characteristics of populations of organisms over time. By studying fossils and living organisms, scientists have discovered that many of today's organisms are very different from those that lived on Earth millions of years ago. Most biologists believe these differences result from evolution.

Most biologists believe evolution occurs when random changes that make an organism better adapted to its ENVIRONMENT occur in the genetic material of the organism. Such changes help the organism survive and reproduce. When the organism reproduces, it passes these traits to its offspring. Thus, the offspring, like its parent, has traits that make it well suited to its environment. Through reproduction, these traits are passed to future generations. Over many generations, enough changes may occur in a population to produce a new SPECIES.

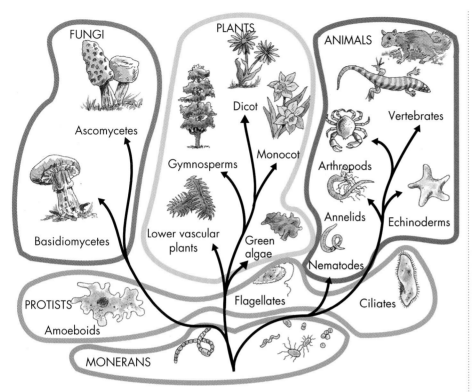

FUNGI

Ascomycetes

Basidiomycetes

PLANTS

Dicot

Gymnosperms

Monocot

Lower vascular plants

Green algae

Flagellates

ANIMALS

Vertebrates

Arthropods

Annelids

Echinoderms

Nematodes

Ciliates

PROTISTS

Amoeboids

MONERANS

◆ This diagram is called an *evolutionary tree*. It shows one possible sequence in the evolutionary development of all of today's major groups of organisms from a single species that lived on Earth millions of years ago.

Natural Selection

The term *natural selection* is sometimes referred to as "survival of the fittest." In a population, some organisms have traits that make them better adapted to their environment than other members of their species. For example, a strong and fast-moving PREDATOR such as a shark is more likely to capture the prey it uses for food than is a weaker, slower member of its species. As a result of this ability, the stronger, faster-moving predator is better adapted to its way of life and more likely to survive in its environment (the open OCEAN) than is the weaker, slower predator. Thus, the most fit organism survives, while less fit members of the population do not.

The members of a population that survive and reproduce pass the traits that made them successful in their environment to their offspring. For example, a fast-moving shark that is able to get plenty of food is more likely to survive into adulthood and reproduce than are slower members of its species. As it reproduces, some of the traits that made the shark a fast swimmer and successful hunter will be passed to its offspring. These traits will help the offspring be successful swimmers and hunters, survive into adulthood, and reproduce to pass these traits to their offspring. Over many generations, members of the species change and become better adapted to their environment than were their ancestors. In this way, nature favors, or selects, the survival of some members of a population over others.

PHOTOSYNTHESIS. Because the leaves and branches of the trees block out much of the sunlight from reaching the forest floor, the plants, such as mosses and FERNS, that commonly live on the forest floor are often adapted to live in areas of shade. Other plants that grow near ground level and need much sunlight to survive are seldom present in a forest. Besides competing for sunlight, plants living on a forest floor compete with trees for water and living space. Organisms successful at getting the resources they need to live are likely to survive into adulthood and reproduce. The unsuccessful organisms die.

Variations

All species show some variations among the members of a population. Variations are differences in the traits that are among organisms of the same species. In humans, variations include such observable differences as height, weight, and eye color. Such variations may result from the unique combination of GENES that combine during sexual reproduction. These variations may also result from mutations in the genes or chromosomes of organisms. In natural populations, variations may either help or harm a given organism's chances for survival.

EVOLUTION OF SPECIES

Through the process of evolution, the traits of populations change from generation to generation. Over time, in some cases millions of years, such changes lead to the evolution of new species. Currently, scientists have identified more than 1.5 million different species of organisms, which they have classified into five major groups, called *kingdoms.* The kingdoms include the monerans, or BACTERIA; the protists; the FUNGI; the plants; and the animals. [*See also* ADAPTATION; ADAPTIVE RADIATION; BIODIVERSITY; CARRYING CAPACITY; COEVOLUTION; CONVERGENT EVOLUTION; GENE POOL; GENETIC DIVERSITY; GENETICS; HABITAT; MASS EXTINCTION; NONRENEWABLE RESOURCES; OVERPOPULATION; POPULATION GROWTH; RENEWABLE RESOURCES; SPECIES DIVERSITY; and SUBSPECIES.]

Exotic Species

❚A type of organism that is not native to a particular region. Exotic species are sometimes called *alien, nonnative,* or *introduced species.* The SPECIES that naturally inhabit a particular area or ECOSYSTEM are called NATIVE SPECIES. Such species are adapted to survival in their ecosystems. Often, through accidental or deliberate means, a new species is introduced to an area. Such a species is called an *exotic species.*

◆ The mongoose was introduced into the Hawaiian Islands, where it devastated the bird population. It is a major cause of the extinction of many native Hawaiian species.

Many organisms people are familiar with are exotic species. For example, cats, rats, and dogs are examples of exotic species. Each of these species evolved in a particular region of Earth. However, through the activities of people over thousands of years, each of these organisms has become common in virtually all of the nations of the world.

The arrival of exotic species to an ecosystem may occur by accident. For example, a species of mussel called the zebra mussel has been introduced into the GREAT LAKES region of the United States. The species, which is not native to the United States, is believed to have been carried to the area aboard a cargo ship that passed through the Great Lakes. Since its arrival, the mussel has reproduced in great numbers, outcompeting the native species for the NATURAL RESOURCES it needs for its survival. In addition, the mussel has migrated into various drainage pipes throughout the region. In the pipes,

the mussel has reproduced in such numbers that the pipes have become completely blocked and unable to carry liquids. Locating the sources of these blockages and removing the mussel causing the problem has cost the people living in the area millions of dollars.

People often introduce exotic species into an area intentionally. In some cases the species are brought to an area to feed on other organisms that are considered pests. In other cases, species may be introduced to an area because people find the species attractive. Such was the case with the water hyacinth, a flower native to South America that was introduced to the waters of Florida. Since its introduction, the water hyacinth has outcompeted many other water PLANTS and thrived. Today, it is estimated that the water hyacinth has invaded more than 1,850,000 acres (740,000 hectares) of lake and river waters across the southern portion of the United States.

◆ The water hyacinth, native to South America, was introduced to Florida, where it grew excessively. Today, the abundant water hyacinth clogs several locks and canals.

Introduced or exotic species often have no PREDATORS, parasites, or diseases to which they are vulnerable in their new ecosystem. Without these factors, the exotic species population can grow quickly and almost unchecked. This rapid POPULATION GROWTH aids the new species in its survival and helps to outcompete native species for resources.

Exotic species may threaten the survival of the native species in other ways. For example, the new species must compete with the native species for resources such as food, water, and living space. Often, the new species becomes the predator or prey of native species. In addition, a new species may bring to the ecosystem diseases to which the native species have no natural defenses. This is especially problematic in island HABITATS where native plants and animals have not evolved many defenses. For example, introduced species brought to the Hawaiian Islands a form of malaria that infects BIRDS. Many native birds, lacking defenses from this disease, died of the avian malaria. The bird population was further devastated through the introduction of the exotic mongoose to the islands. The mongoose is a predatory animal that often feeds on birds. The mongoose has been a major cause of the EXTINCTION or near extinction of many native Hawaiian species. All of these factors may disrupt the natural balance within the ecosystem and threaten the survival of native species. [*See also* BIODIVERSITY; BIOLOGICAL CONTROL; BIOREGION; CARRYING CAPACITY; CLIMAX COMMUNITY; COMPETITION; CONSERVATION; ENDANGERED SPECIES; HABITAT LOSS; INTEGRATED PEST MANAGEMENT (IPM); and KEYSTONE SPECIES.]

Exponential Growth

A pattern of rapid POPULATION GROWTH in which the number of individuals increases by a constant percentage each generation, resulting in a population explosion. Exponential growth is very different from more familiar types of growth. Consider, for instance, the growth of a paycheck. Suppose a person earns $100 for every week worked. If the person works two weeks, he or she receives $200. If he or she works three weeks, he or she earns $300. This type of growth is called *linear,* or *arithmetic, growth* because it occurs at a steady, unchanging rate. When graphed, linear growth appears as a straight line.

Exponential growth is different from linear growth and shows a greater increase in numbers. When graphed, exponential growth appears as a J-shaped curve. Consider, for example, a population of organisms growing at its maximum growth rate. This would be a population in which every individual survived to adulthood and produced offspring. If the population begins with only two individuals, growth would be slow at first. However, with each passing generation, an increasing number of individuals would be added to the population. For example, the first generation would produce two individuals. The second would produce four. The third would produce eight, and so on. Unlike linear growth, the growth rate in a population showing exponential growth is changing

each generation. Under ideal conditions—unlimited food and water, lack of PREDATORS, and absence of disease—exponential growth can permit a population of organisms to explode in size in a very short time. In fact, as long as a population's birth rate is greater than its death rate, it will continue to grow in size.

Populations never grow at their maximum growth rates. Rather, population size is limited by many factors in the ENVIRONMENT. For animals, limiting factors might include availability of a food source, water, or shelter. For PLANTS, the supply of water, sunlight, or a particular mineral nutrient may be the limiting resource. If the population increases, more and more individuals must use the limiting resources. Thus, it becomes harder and harder for individuals to survive. Members of the population begin to compete with each other and with other populations for the resources. As COMPETITION increases, the death

rate rises. At the same time, the birth rate declines. Eventually, the population reaches a size at which it ceases to grow.

When a population grows to the point where the environment is no longer able to support it, growth stops and begins to level out. At this point, the CARRYING CAPACITY of the environment has been reached. The carrying capacity is the maximum number of individuals an environment can support over a long period of time. Sometimes a population can reach its carrying capacity and, through its own actions or a shift in the environment, be drastically reduced in a population crash. A population of deer reaching the carrying capacity of available food plants, for instance, might crash with the onset of an especially harsh winter. As the food supply increases, the population of deer would begin its exponential growth again. [*See also* BIOLOGICAL COMMUNITY; DEMOGRAPHY; LIMITS TO GROWTH; and POPULATION GROWTH.]

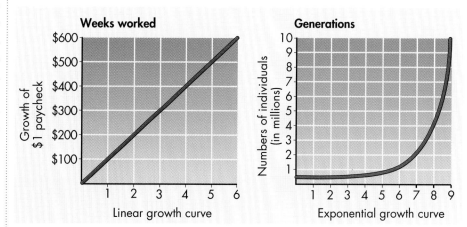

◆ Exponential growth is very different from linear growth. If there are no limits to the growth of a population, for example, it will grow exponentially. Each generation will contain a larger number of individuals than the one before it.

◆ The mammoth is an extinct relative of the modern elephant.

Extinction

▶The complete disappearance of SPECIES from Earth. Often, people refer to local extinctions, or extirpations. For example, timber wolves are still found commonly in Alaska, much of Canada, and a few northern states, but they are extinct in places like New Jersey and Pennsylvania, where they were once found. Extinction occurs when a species cannot adapt to a changed environment

and/or cannot reproduce successfully under changed conditions. The change can be natural, as in a change in CLIAMTE, or the result of a disturbance of natural systems by humans. A species can disappear quickly, as a result of some catastrophe, or slowly.

Because we only know of a portion of prehistoric organisms from fossils, no one knows how many species have ever existed on Earth. Some researchers estimate that there have been at least 450 million species since life began. Almost all of these are now extinct.

In the history of Earth, there

have been several MASS EXTINCTIONS, when great numbers of species became extinct at about the same time. The last mass extinction, and the most famous, occurred about 65 million years ago, when all of the dinosaurs, most marine reptiles, all flying reptiles, marine INVERTEBRATES, and other marine species died. The most widely accepted theory for the cause of this extinction was the impact of a large asteroid. After a mass extinction, it can take millions of years for the survivors to give rise to new species that fill in the NICHES in the affected HABITATS.

Species can become extinct for many reasons. Habitat can be altered by sudden natural changes or through the actions of humans. For example, one reason for the extinction of the PASSENGER PIGEONS was the cutting of the FORESTS that served as their nesting and feeding grounds.

COMPETITION with other species also can lead to extinction. For example, when South America was an island, it was inhabited by a great number of large carnivorous flightless birds. When North and South America became connected, large predatory mammals like cougars and coyotes crossed into South America and competed with or consumed the large birds, driving them to extinction.

◆ By studying fossils, such as this ammonite fossil, scientists can learn about organisms that have become extinct.

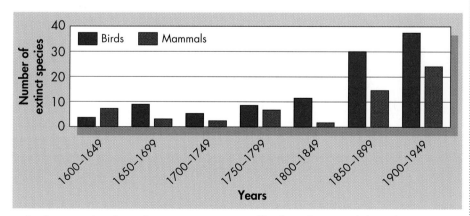

◆ As shown in the chart above, many species of birds and mammals have become extinct since the 1600s, with the greatest number of extinctions occurring in the nineteenth and twentieth centuries.

Some species have been hunted to extinction. An example is the dodo bird of the island of Mauritius, killed off by European settlers. Many researchers suspect that some of the large animals that survived past the end of the last ice age, like mammoths and mastodons, were hunted to extinction by expanding populations of people.

Pollution also can drive a species to extinction. In the 1800s, one of the most common INSECTS in the eastern United States was the American burying beetle, which feeds on animal carcasses, helping to break them down in the SOIL. This beetle, however, has almost completely vanished. It is thought that the main reason for the beetle's decline has been the use of PESTICIDES.

Currently, directly and indirectly, humans are causing a huge wave of extinctions. Through HABITAT DESTRUCTION, the introduction of EXOTIC SPECIES, POLLUTION, unregulated HUNTING, and POACHING, people put several stresses on species, forcing them to contend with increased predation, competition, pollution, and habitat destruction all at once. When a population greatly decreases in size, small disasters, or problems that arise from a lack of genetic dervisity, can easily finish off that species.

Species of animals, plants, and INVERTEBRATES are vanishing throughout the world. In species-rich areas, such as the tropical RAIN FOREST, some species are being destroyed before they even can be described and named. Some researchers estimate that human activity causes extinctions at an approximate rate of 17,500 species per year. Although most of these are plants and insects, all are important parts of biodiversity.

Attempts are being made to slow the current rates of extinction. Laws such as the ENDANGERED SPECIES ACT and organizations such as the CONVENTION ON INTERNATIONAL TRADE IN ENDANGERED SPECIES OF WILD FAUNA AND FLORA (CITES) are important steps in the protection of SPECIES DIVERSITY. Once a species becomes extinct, it is lost forever and with it goes any hope of benefits we might have gained from its role in the ECOSYSTEM. [*See also* ENDANGERED SPECIES, EVOLUTION, HABITAT LOSS, RESTORATION BIOLOGY, WILDLIFE CONSERVATION and WILDLIFE MANAGEMENT.]

Exxon Valdez

▌The oil tanker involved in the worst OIL SPILL in U.S. history. Many people think that the *Exxon Valdez* oil spill and the poor results of the effort to clean it up was proof that oil companies are too careless in their treatment of the ENVIRONMENT. The *Exxon Valdez* was a supertanker owned by the Exxon oil company. The tanker operated off Valdez, Alaska, at the endpoint of the ALASKA PIPELINE. At four minutes past midnight on March 24, 1989, the *Exxon Valdez* ran aground on a reef in Prince William Sound, about 25 miles (40 kilometers) from Valdez. The tanker spilled more than 11 million gallons (41 million liters) of oil into the sea.

Prince William Sound is the HABITAT of a great variety of WILDLIFE, including FISH, BIRDS, and marine MAMMALS such as seals and sea otters. Along its shores are bird-nesting areas, salmon-hatching sites, local fishing businesses, as well as the Katmai National Park.

Before the *Exxon Valdez* spill, some spokespeople for the oil industry said that it was prepared to respond quickly in an emergency and could contain any oil spill within five hours. However, very little oil was recovered from Prince William Sound after the spill. Local citizens and Alaskan state officials finally started a cleanup effort themselves.

Then the weather got worse and the oil slick became impossible to control. It polluted hundreds of miles of Alaskan shoreline, including the shore at Katmai National Park. It killed countless numbers of fish and shellfish, at least 50,000 birds, and hundreds of other animals. Coastal communities that depended on fishing were severely affected.

The oil company and others tried to clean up the shores throughout the summer of 1989, but much of the damage could not be undone. On many beaches, the oil had soaked too deep into the ground to remove. Some beaches that had been cleaned up were polluted again when more oil washed in from the sea. Two years after the spill, the NATIONAL OCEANIC AND ATMOSPHERIC ADMINISTRATION (NOAA) concluded that the damage was even greater than earlier reports had indicated.

Exxon became the target of much public criticism and many lawsuits. Some people boycotted Exxon gasoline. In the end, the oil spill resulted in major regulatory changes, improvement of cleanup methods, and increased public awareness. [*See also* MARINE POLLUTION; OIL DRILLING; OIL POLLUTION; and OIL SPILLS.]

◆ The worst oil spill in U.S. history occurred when the *Exxon Valdez* ran aground on a reef in Prince William Sound in Alaska.

◆ Sea otters were among the wildlife affected by the spill.

◆ Eleven million gallons (41 million liters) of oil were spilled into the sea by the *Exxon Valdez.*

Anchorage

Valdez

Cook Inlet

Cordova

Homer

Seward

Prince Williams Sound

Montague Island

Gore Point

Kodiak

Gulf of Alaska

Alaskan Peninsula

Kodiak Island

Kodiak National Wildlife Refuge

Cumulative extent of the oil spill

National Wildlife Reserves

◆ Area affected by the *Exxon Valdez* oil spill. Much of the coast of Prince William Sound is breeding habitat for wildlife.

F

Family Planning

▌▶Practice in which people decide when to have children and how many they will have. The size of the human population is growing at a rapid rate. This POPULATION GROWTH places great demands on the ENVIRONMENT, especially on its NONRENEWABLE RESOURCES and energy resources. To ensure that the human population does not exceed Earth's CARRYING CAPACITY for humans, many people emphasize the need for population control.

Family planning is one method by which the growth of the human population can be controlled. Many population experts believe that family planning is the only way people can control human population growth.

For family planning to be an effective means of population control, several conditions must be met. First, people must be provided with information about methods of **birth control**. Second, birth control devices and medications must be available to people. In addition, people must have a desire to practice birth control.

People become involved in family planning for a variety of reasons. In many cases, family planning is voluntary. People participate in family planning to make sure they have the resources, such as a home, maturity, and financial

◆ In many cities worldwide, family planning centers are established to provide the public with information on birth control methods.

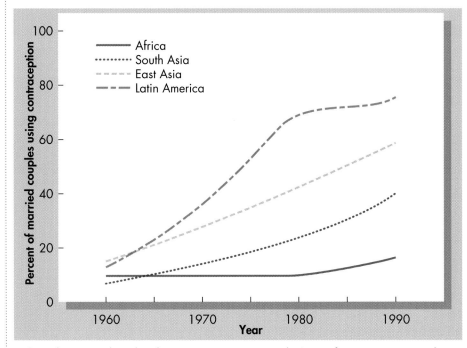

◆ Over the years, there has been an increase in couples' use of contraception to plan their families.

security, to raise children in a safe and happy environment. In some regions of the world, however, family planning is required by the government. For example, the governments of both China and India have passed laws to restrict the number of children a couple may have. These laws are designed to prevent the populations of these countries from growing out of control.

Countries that have laws controlling the number of children in a family may have to enforce their laws in a variety of ways. In some countries, families face penalties, such as rations on food and other resources, if they have more children than the government recommends. Other countries require families that have too many children to pay additional taxes as a penalty. In extreme cases, a government may require the **sterilization** of a couple that has more than the designated number of children. [*See also* AGE STRUCTURE; and DEMOGRAPHY.]

Famine

A severe food shortage that results in the MALNUTRITION or starvation of many people. In some parts of the world, famine has been a significant cause of death. Deaths resulting from famine are most often the result of starvation—people do not have enough food to survive. Famine may also lead to death from malnutrition—the lack of one or more nutrients needed by the body. The food shortages that result in famine have a variety of specific causes, which can be classified into two broad groupings: natural events and human-caused problems.

A drought—an extended time period with little or no rainfall—is a weather condition that can lead to famine. During a drought, crops cannot grow because of the shortage of water. Unlike a drought, which occurs over a period of time, severe storms, floods, volcanic eruptions, and fires are NATURAL DISASTERS that can quickly lead to famine. Each of these conditions can devastate an area, destroying both crops and LIVESTOCK.

Sometimes organisms such as INSECTS, BACTERIA, and FUNGI can destroy crops to such an extent that a famine results. This is particularly true in places in which the people rely mainly on one food to meet their nutritional needs. For instance, in the 1840s in Ireland, the primary food crop was the potato. A water mold, a type of fungus that is a parasite of the potato plant, destroyed much of the potato crop several years in a row. Because potatoes were such an important part of the diet in Ireland, many

◆ In parts of Africa, food is rationed to famine victims.

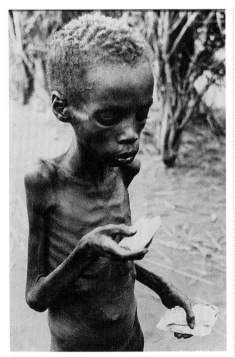

In some parts of the world, relief centers are established to distribute food to people.

Malnutrition is a significant cause of death in some countries.

people died of starvation or fled the country to avoid starvation.

The social and political activities of people, especially wars, have also caused famines. The famines that plagued Ethiopia in the late 1980s and Somalia in the early 1990s both resulted largely from conflicts over control of the government and resources. In other cases, famines have resulted from poor farming methods that lead to such problems as soil EROSION—the carrying away of TOPSOIL—and DESERTIFICATION—the changing of land in a semiarid region into a DESERT as a result of improper use of the land. Each of these conditions can leave an area unsuitable for the growth of crops. [*See also* DEFORESTATION; HEALTH AND NUTRITION; OVERGRAZING; and SALINIZATION.]

Federal Energy Regulatory Commission (FERC)

◗A five-member committee within the U.S. Department of Energy responsible for setting prices of NATURAL GAS and ELECTRICITY. The commission is also responsible for licensing power projects and approving all oil and gas pipelines.

The Federal Energy Regulatory Commission (FERC) replaced the Federal Power Commission (FPC) in 1977. The original purpose of the FPC when it was created in the 1920s was to manage and issue licenses to HYDROELECTRIC POWER plants in the United States. Later, the FPC was given the responsibility for setting prices on the sale of natural gas and electricity.

In 1977, Congress passed the Department of Energy Reorganization Act, which authorized the government to restructure its energy policies. The FERC was created and given the job of handling all of the old functions of the FPC, as well as new responsibilities.

These new functions include setting rates for the transportation of oil by pipelines and determining the value of the pipelines themselves. [*See also* FOSSIL FUEL.]

Federal Insecticide, Fungicide, and Rodenticide Act (FIFRA)

▮Law that gives the ENVIRONMENTAL PROTECTION AGENCY (EPA) the power to regulate the use of PESTICIDES. Pesticides are chemical substances that are used to kill unwanted organisms. Examples include INSECTICIDES, FUNGICIDES, RODENTICIDES, and HERBICIDES. Insecticides kill INSECTS on farms, in homes, and other places. Fungicides prevent fungal diseases that contaminate food crops by killing the FUNGI that cause disease. Fungicides are also components of disinfectants used in homes and hospitals. Rodenticides protect stored food from rats and mice. Approximately 50,000 pesticides are registered in the United States. Some of the 1,500 different active ingredients in the pesticides can be harmful to people and WILDLIFE.

The Federal Insecticide, Fungicide, and Rodenticide Act (FIFRA) was originally passed in 1947 and strengthened in 1972. Under the act, the EPA can:

• establish strict conditions for the registration of all new pesticides;

◆ Warnings must be posted to protect the public when food containing rodenticide is placed on the ground.

• ban or restrict the use of pesticides already in use, if they are deemed too risky to human health or the ENVIRONMENT; and

• set standards for how pesticides will be applied and how containers will be labeled.

FIFRA gives individual states the power to adopt even stricter standards if they choose. For example, a state can ban a pesticide even if the EPA has not done so.

A state can also enforce tougher right-to-know laws for workers and consumers, so that people are educated about the possible risks of a pesticide. Many states enacted laws to reduce pesticide contamination when the poisons began to show up in water supplies. [*See also* DDT; DIOXIN; LAW, ENVIRONMENTAL; and PEST CONTROL.]

Federal Mining Act

▮A law enacted in 1872 that regulates the MINING of materials on PUBLIC LANDS. The Federal Mining Act of 1872 was signed into law by President Ulysses S. Grant. This law, which is still in effect today, regulates the mining of hard-rock MINERALS, such as silver, gold, and other metals, on public lands. When it was established, the primary goal of the Federal Mining Act was to promote settlement in the

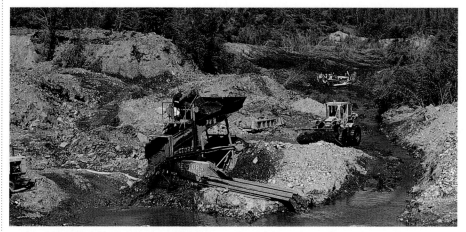

◆ The Federal Mining Act of 1872 allows individuals to use and even purchase public lands at very inexpensive prices for the removal of the minerals the land contains.

western United States by giving individuals and small companies the right to stake a mining claim on public lands. According to the law, an individual can remove metal-containing minerals from public lands without paying federal taxes on any of the profits. The law also permits the miner to purchase the land from the federal government for less than $5 per acre. In doing so, the miner gains all mining rights to the land and at the same time ensures that other individuals cannot remove minerals from it.

Since 1872, more than 3.5 million acres (1.4 million hectares) of public land have been sold to individuals and companies, both American and foreign, at the low price established in 1872. It is estimated that these lands contain almost $100 billion worth of minerals. However, the federal government and people of the United States do not share in the profits made from these minerals, because of the provisions of the Federal Mining Act. In addition, some of these once-public lands have been developed by their owners for housing or other uses.

At the time the Federal Mining Act was created, people knew little about and were not concerned with the consequences of mining on the ENVIRONMENT. Therefore, no provisions were included for reclaiming land that is no longer used for mining. Fortunately, many states have enacted such requirements and are now making efforts to restore land that has been damaged by mining practices. [*See also* OPEN-PIT MINING; POLLUTION; RESTORATION BIOLOGY; RECLAMATION ACT; SURFACE MINING; TAILINGS; and TOXIC WASTE.]

Fern

▶ A nonflowering PLANT that reproduces from spores rather than seeds. Ferns are characterized by their leaves, called *fronds,* which often have a delicate, lacy appearance. *Spores,* the reproductive cells of ferns, form on the fronds. There are approximately 10,000 living SPECIES of ferns. Most ferns are only a few feet high. However, several tree-sized fern species live in the TROPICS.

Ferns live in a variety of HABITATS, including swamps, dry FORESTS, and the sides of rocky cliffs. Some fern species are very hardy and capable of withstanding harsh conditions. However, all species of ferns require moisture in the ENVIRONMENT in order to reproduce.

Ferns, along with club mosses and horsetails, are the simplest types of vascular plants. Vascular plants have tubelike tissues that carry water, nutrients, and food produced by PHOTOSYNTHESIS throughout the plant. In some plants, vascular tissue can carry materials over long distances, from roots that are buried deep underground to treetops more than 163 feet (50 meters) above the ground. Because of these adaptations, vascular plants can live in a wide variety of habitats and grow to large sizes.

Ferns and other simple vascular plants first appeared in the fossil record nearly 400 million years ago. The earliest ferns were tall and treelike, forming vast forests in warm, swampy environments. The forests

◆ Ferns are a simple type of vascular plant. Ancient fern forests provide the coal used today.

characterized Earth's Carboniferous period and are the source of the PEAT and COAL used by people today.

About 280 million years ago, when ferns and other simple plants had reached their greatest numbers and SPECIES DIVERSITY, Earth's CLIMATE began to change. As the fern forests died and were replaced by succeeding generations of plants, thick layers of plant material began to collect in the ancient swamps. Over long periods of time, the swamps drained and filled again. Layers of plant material and SEDIMENTS began pressing down on one another. Eventually, this combination of plant material and pressure produced coal—the shiny, hard, black material that is burned today in coal-fired furnaces to produce heat and ELECTRICITY. [*See also* CARBON; FOSSIL FUELS; and NONRENEWABLE RESOURCES.]

Fertility Rate

❿The average number of babies born to women of childbearing age, as indicated by population studies. The fertility rates of a population provide data that are important to an understanding of general trends in POPULATION GROWTH. Such information is used to determine how the population needs of a society or community may change over time. To determine a fertility rate, accurate measurements of the number of births that occur in a population must be made.

There are several methods used to calculate fertility rates. The *general fertility rate* of a population indicates the total number of babies born in one year for each 1,000 women of childbearing age. The general fertility rate is often used to show the number of births in a population, but it does not provide information about the average age at which the women in the population sample begin having children. The general fertility rate also does not indicate at what age women stop having children.

Women who begin bearing children at a very young age are more likely to have many children than are women who begin families at an older age. This trend may affect a country's actual population growth. Thus, many demographers, people who study populations, calculate the growth of a population using the *age-specific fertility rate.* An age-specific fertility rate takes the AGE STRUCTURE of a population into consideration. Data are then calculated as the number of births per year per 1,000 women in a specific age group. Such a calculation helps to provide information about what the future needs of a society will be, as well as data about when those needs will have to be met.

Fertility data can also be calculated as the *total fertility rate.* The total fertility rate is the number of children an average woman bears in her lifetime. This figure is based on a hypothetical woman and on the general fertility rate for a population.

Finally, some demographers calculate data about fertility rates in terms of the *replacement fertility rate.* The replacement fertility rate is the number of children each couple must have to replace themselves and thus maintain population at a constant level. The minimum replacement fertility rate for a couple is two offspring. However, the actual figure is likely to be greater than two children, because not all children survive. Thus, if the replacement fertility rate is too low, a country's population may actually decline in size. [*See also* DEMOGRAPHY and INFANT MORTALITY.]

Fire Ecology

❿The study of fire as a vital ecological factor in maintaining the health of certain ECOSYSTEMS. Fire is a normal fact of life in many ecosystems. Therefore, the native PLANTS and animals have adapted to it and may even require it to survive.

People think of fire as dangerous. For many years ecologists and land managers sought to prevent fires from occurring in all ecosystems. Experience taught them, though, that the lack of fire can sometimes turn out to be detrimental, not beneficial. Today, fire ecologists learn about the best way to allow fires to burn in certain ecosystems, using the practice of PRESCRIBED BURN.

Particular plants that dominate certain ecosystems are especially dependent on fire for their survival. Fire-tolerant plants, such as the grasses of the plains, concentrate

◆ Periodic fires help prairie grasses regenerate and enrich the soil.

◆ It will be years before new trees reach the height of old ones that were burned in a fire in Yellowstone National Park. This photo was taken two years after the fire of 1988.

their energy on building extensive root systems that survive fire and that sprout quickly after the above-ground parts of the plant have burned. Some SPECIES of pine tree are fire dependent. These pines put most of their energy into producing lots of seeds—seeds that need the scourge of fire to germinate. The mature trees may die in the fire, but their numerous seeds will live on.

Fire is crucial to the maintenance of GRASSLANDS. Grasses quickly sprout fresh new shoots after a fire. Trees and bushes are much slower to recover from fire damage. Fire thus prevents trees and shrubs from taking over grasslands and ensures the dominance of grasses on the PRAIRIE. On the prairie, most fires start when lightning strikes the ground and sets it ablaze.

In some FORESTS of the southeastern United States, fire destroys hardwoods and other tree species that are less fire tolerant than the long-leaf pine. Thus, because of fire, this pine dominates these southern forests and supports a unique forest ecosystem.

A lightning bolt strikes somewhere on Earth every minute of every day. Natural fires occur in all ecosystems at some time and, to a certain degree, have a positive effect on most ecosystems.

In the summer of 1988, a lightning bolt started a fire in YELLOWSTONE NATIONAL PARK in Wyoming. The policy of the NATIONAL PARK SERVICE was to let a fire caused by lightning burn unless it threatened property, scenic areas, or ENDANGERED SPECIES. Up until 1988, about 200 naturally caused fires had been allowed to burn in Yellowstone. But the summer of 1988 was different because the worst drought ever recorded was plaguing the region. Several natural fires had been burning for nearly a month when, in July, some hikers caused what turned out to be the worst fire of all. Though about 25,000 fire fighters were called in to try to control the fire, vast tracts of the park were severely damaged.

Some politicians blamed the Park Service for its "let natural fires burn" policy. Scientists, however, defended the Park Service but suggested that perhaps that policy did not go far enough. They believed that, over the years, too many fires had been put out. The scientists pointed out that if small fires are not allowed to burn, organic matter, such as fallen leaves, twigs, and branches, begin to accumulate on the forest floor. In dry weather, this accumulation of debris adds fuel to any fires that do occur, making those fires much worse.

In many ecosystems, naturally caused fires are like brooms that sweep the ground clean of excessive debris. If these fires occur periodically when there is relatively little debris on the ground, the fire will not have much fuel and will probably not be too destructive. But if small fires are put out before they can accomplish their "housecleaning" work, years' worth of debris will feed the fire that inevitably comes and may turn it into a devastating firestorm.

Fish

▶Streamlined, aquatic animal that breathes with gills and moves about with fins. Fish make up one of the five subgroups of VERTEBRATES, animals with a backbone and internal

◆ A shark is a cartilaginous fish; its skeleton is made of cartilage rather than bone.

skeletons made of bone or **cartilage**. In addition to fish, vertebrates include AMPHIBIANS, BIRDS, MAMMALS, and REPTILES.

DIVERSITY OF FISH

Fish inhabit nearly every type of aquatic HABITAT on Earth—lakes, ponds, streams, rivers, OCEANS, and WETLANDS. They are adapted to living in shallow, warm waters, as well as in deep, cool waters. Some fish can even survive in heavily polluted water.

With nearly 20,000 different SPECIES, fish are the most abundant and diverse vertebrates. They range in size from the tiny dwarf goby, which is less than 0.4 inch (1 centimeter) in length, to the huge whale shark, which reaches lengths of up to 50 feet (15 meters). Fish

THE LANGUAGE OF THE ENVIRONMENT

cartilage soft, elastic body tissue; in humans, cartilage can be found in the nose, ears, and body joints.

cartilaginous made of cartilage.

tissues groups of cells that work together to carry out a function; examples include skin tissue, muscle tissue, and nerve tissue.

also exhibit an assortment of unusual and interesting shapes, from the S-shaped seahorses to the flattened, kite-shaped skates and rays.

Biologists recognize three major groups of fish: (1) jawless fish, which include the long, tubelike lampreys and hagfish; (2) **cartilaginous** fish, with skeletons composed of cartilage rather than bone, such as sharks, skates, and rays; and (3) bony fish, which include TUNA, SALMON, goldfish, and many other familiar fish.

BIOLOGY OF FISH

Although fish are an extremely diverse group, they do share certain characteristics, including the ADAPTATIONS that allow them to survive

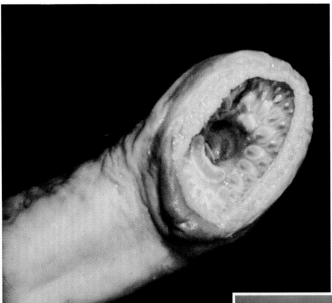

Fish and Wildlife Service

◆ Fish come in a great variety of colors and shapes.

in aquatic habitats. The most obvious adaptations are those on the outside of a fish's body. Most fish have sleek, streamlined shapes that allow them to glide effortlessly through water. Fish also have paired fins and bladelike tails, which help them turn and push through the water.

Scales are another adaptation common to fish. Scales are thin plates of bone formed by the skin of the fish. Different species of fish have scales of different shapes and colors. Scientists can often identify fish by examining only the scales. For instance, tooth-shaped scales are characteristic of sharks, whereas cone-shaped or round scales are characteristic of salmon.

Gills are another characteristic that can be clearly seen on the outside of a fish's body. All fish have gills, which allow them to obtain DISSOLVED OXYGEN from the water. Gills are thin **tissues** that have many blood vessels. When water

passes over the gills, the blood vessels absorb the dissolved oxygen. The oxygen is then circulated to the cells of the body. [*See also* CORAL REEF; DDT; FISHING, COMMERCIAL; FISHING, RECREATIONAL; FISH LADDER; and OCEAN.]

▶An agency within the U.S. DEPARTMENT OF THE INTERIOR that is responsible for the CONSERVATION of the BIRDS, MAMMALS, FISH, and other WILDLIFE of the United States. It is also responsible for the protection of ENDANGERED SPECIES and the restoration of wildlife HABITAT.

The Fish and Wildlife Service became an agency of the U.S. Department of the Interior in 1940. It was formed when the Bureau of Fisheries (created in 1871) and the Bureau of Biological Wildlife Survey (created in 1885) were combined to form a single agency. Since its creation, the Fish and Wildlife Service has been held responsible for the three main tasks of conserving wildlife, operating breeding programs for wildlife, and regulating activities related to wildlife.

To meet its goals, the Fish and Wildlife Service is responsible for managing 150 areas used by waterfowl and 442 wildlife refuges. These

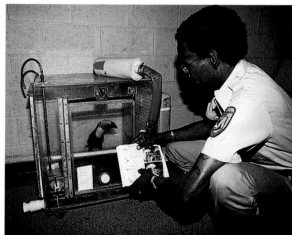

◆ The Fish and Wildlife Service regulates activities related to wildlife, including the trade in exotic birds.

areas cover more than 90 million acres (36 million hectares) of land in the United States. In addition, the Fish and Wildlife Service maintains several research facilities and about 70 fish hatcheries throughout the United States. The research facilities are often involved in CAPTIVE PROPAGATION programs which are designed to protect endangered species from EXTINCTION. The hatcheries of the Fish and Wildlife Service are used for raising fishes that are later released into stock ponds, lakes, and streams throughout the country.

Among its many jobs, the Fish and Wildlife Service may be best known for developing and maintaining an endangered species list. In this role, the agency is responsible for identifying those SPECIES that are most at risk of extinction, as well as the major threat to each one's survival. The agency then develops or enforces regulations designed to protect the species. It may become involved in habitat restoration projects to increase its

chances of survival in the wild. In addition, the agency must gather and update data about the approximate population size of each species it identifies as threatened or endangered. [*See also* ARCTIC NATIONAL WILDLIFE REFUGE (ANWR); BIODIVERSITY; CONVENTION ON INTERNATIONAL TRADE IN ENDANGERED SPECIES OF WILD FAUNA AND FLORA (CITES); CORAL REEF; ENDANGERED SPECIES ACT; ENVIRONMENTAL IMPACT STATEMENT; ESTUARY; EXOTIC SPECIES; FISHING, RECREATIONAL; FOSSEY, DIAN; HABITAT LOSS; HUNTING; INTERNATIONAL UNION FOR THE CONSERVATION OF NATURE (IUCN); LAW, ENVIRONMENTAL; MARINE MAMMAL PROTECTION ACT; MARINE PROTECTION, RESEARCH AND SANCTUARIES ACT; NATIONAL ENVIRONMENTAL POLICY ACT (NEPA); NATIONAL WILDLIFE REFUGE; OLD-GROWTH FOREST; WETLANDS; WETLANDS PROTECTION ACT; WILDERNESS ACT; WILDLIFE MANAGEMENT; and WILDLIFE REHABILITATION.]

Fishing

See FISHING, COMMERCIAL and FISHING, RECREATIONAL

Fishing, Commercial

▍An industry that harvests, processes, and markets freshwater and marine animals. Commercial fishing provides an important source of protein for the world's population, as well as fish by-products for use in industry and agriculture. Unlike most methods of food production, commercial fishing usually involves taking animals from freely moving wild populations. This practice leads to special challenges in the management of these populations and in the regulation of the industry.

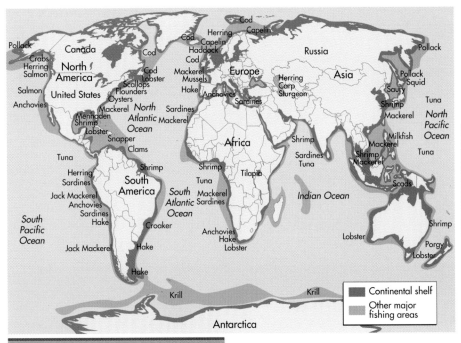

◆ Fish are most abundant just offshore, in the relatively shallow, sunny, nutrient-rich water on the continental shelves.

MALS such as WHALES. Because of the large number of species involved, commercial fishing methods in different parts of the world vary. However, most commercial fishing involves one or more of the following techniques: impounding, entangling, or using hooks and lines. All of these methods can be assisted by **sonar** devices used for locating animals underwater, radar devices that improve navigation, and facilities on the ship that are used to process the catch.

Impounding involves trapping. It may be accomplished by using actual traps or by surrounding schools of fish with long nets such as the *purse seine*. Entangling also involves the use of nets. For example, one method of capturing fish by entangling uses GILL NETS. These

A great variety of aquatic SPECIES are harvested from the world's marine and freshwater environments in addition to FISH. For example, commercial fishers harvest crustaceans such as shrimp, crabs, and lobsters; mollusks such as clams, oysters, and squid; and MAM-

◆ Many coastal countries of the world have commercial fishing industries.

nets are suspended in the water like curtains are. As fish attempt to swim through the net, they become entangled in its fibers. A single fishing net can hold a catch weighing thousands of pounds. Hook-and-line fishing may involve hundreds of hooks set along lines that run for many miles along the ocean bottom.

Fishing is an ancient occupation. For many centuries, commercial fishing was limited by the fact that fish spoil rapidly. Except for some species that could be preserved with salt, fish had to be brought to port, sold immediately, and consumed quickly by local residents. The development of such innovations as canning, refrigeration, and high-speed transportation methods has made it possible to sell fish to more distant markets. This, along with the rapid growth in the world's population and the awareness of the health benefits of eating fish, has resulted in much more intensive **harvesting** of fish resources throughout the world. This is particularly true of countries, like Japan, that have little land for farm-based protein production to feed their large populations. Large-scale commercial fishing is also conducted by Russia, China, Peru, Chile, and the United States. Currently, the U.S. commercial fishing industry is ranked as the sixth largest in the world.

Improved methods for capturing and processing have led to overfishing in some areas. Many aquatic species are being hunted to the point of endangerment, before they can be thoroughly studied. In these cases, it is hard to predict the effects of harvesting, and an unexpected

drop in fish population may occur as well. In addition, because fish swim across national boundaries, it is difficult to form a single plan—agreed to by all countries and private fishing companies—that responsibly manages the fishing industry.

The lack of adequate laws regulating the industry may foster overfishing for short-term economic gain. Several major commercial fisheries have collapsed from overfishing in the twentieth century. Examples of species affected by overfishing include the flounder, cod, and haddock living off the coasts of New England in the United States. To help prevent such problems in the future, CONSERVATION laws like the MARINE MAMMAL PROTECTION ACT and the ENDANGERED SPECIES ACT regulate fishing activities in the United States. Similar laws have been passed by the governments of other countries. [*See also* AQUACULTURE; CORAL REEF; COUSTEAU, JACQUES-YVES; DOLPHINS/PORPOISES; ENDANGERED SPECIES; MARINE PROTECTION, RESEARCH, AND SANCTUARIES ACT; MEAD, SYLVIA EARLE; MERCURY; PLANKTON; SALMON; SEA TURTLE; and TUNA.]

Fishing, Recreational

▶One of the world's most popular outdoor recreational activities, in which people catch FISH for sport. People who use a rod, reel, and line to catch fish are called *anglers*. It is estimated that there are almost

◆ Limits are set to ensure that recreational fishing does not have an adverse effect on the balance of nature in lakes and rivers.

30 million anglers in the United States alone. In just one area, on the seacoast between the states of New York and Virginia, it is estimated that recreational anglers catch three times as many fish as commercial fishers.

TYPES OF FISHING

There are three main types of recreational fishing: game, coarse, and deep-sea fishing. Game anglers fishing in fast-moving streams must cast, or fling, the right bait or lures (artificial bait) on other lines to catch SALMON, bass, trout, and other game fish. Coarse anglers in deep, slow rivers catch fish such as carp and catfish. Deep-sea anglers usually fish for TUNA, marlin, barracuda, shark, swordfish, or other large salt-water fish. People fish in fresh, salt, and brackish (slightly salty) water.

FISHING LAWS

Fishing laws existed in ancient times. Most ancient rulers forbade their subjects to fish in areas reserved for royalty or to take too many fish from other places. In the United States, almost all WILDLIFE is considered the property of the state in which it resides. Each state has an agency to supervise the catching of fish according to local laws.

◆ Fishing is a popular sport in the United States.

Some Fish Facts—and Where Fish Are Caught			
Kind of Fish	**Salt-water**	**Brack-ish**	**Fresh-water**
Bass, Small-mouth	X		
Catfish		X	X
Creek Chub		X	
Flounder	X		
Marlin	X		
Minnow			X
Perch, Yellow		X	
Pickerel			X
Salmon	X		X
Smelt	X		X
Sturgeon	X		X
Trout, Rainbow		X	
Tuna	X		

The state sets the opening and closing of the fishing season—when fishing is allowed—and sells licenses that allow people to fish during the permitted time period. Normally, landowners and their families do not need licenses to fish on their own land, but they must still follow all other fishing regulations.

State fishing laws regulate the manner, size, and number of fish that may be taken. The state also places fish on protected status if too many of the SPECIES have been taken or if POLLUTION has decreased the supply. In ocean waters, officials have authority over fishing up to 3 miles (5 kilometers) from shore. If an angler plans to fish in a new area, he or she must purchase a license and check which fish are protected there. Most states have heavy fines for fishing out of season, exceeding the limit of fish allowed, or taking protected species.

RECREATIONAL BUSINESS

Recreational fishing is also big business for those who sell or rent fishing equipment, including boats. To encourage tourism, some states release fish into recreational fishing grounds to provide ample catch for anglers. Some fishing grounds are set up with special environmental fishing practices—anglers are free to enjoy the sport of catching the fish, but they must then release them back into their natural ENVIRONMENT. [*See also* CONSERVATION; CORAL REEF; ECOSYSTEM; HABITAT; MARINE POLLUTION; SALT MARSH; and WATER POLLUTION.]

◆ Fish ladders help some salmon and other fish make it upstream to spawn, but some fish become too weakened to make the continuous leaps.

Fish Ladder

Angled levels or steps in artificial waterways built to help adult FISH travel up and over DAMS on their way upstream to spawning grounds in the river's headwaters. In the 1930s, when the first dams were built on the Columbia River in Oregon, biologists knew the spawning cycle of SALMON would be interrupted. Every summer, thousands of Pacific salmon instinctively return to the upstream spot where they were spawned, to produce a new generation of fish. In spring, the new generation travels downstream from the freshwater river into the Pacific area. When the dams were built, conservationists and engineers believed that the wild salmon, used to leaping high over waterfalls, could jump over the smaller ladder steps easily. They also felt that salmon bred in hatcheries would make up for any wild salmon that did not make it back to their spawning grounds.

The idea appeared sound, although the ladders were not very successful along the Columbia River. Many adult salmon, weakened by the continuous ladder leaps, died before reaching the spawning grounds. Over the years, tens of millions of hatchlings have been killed by the dams' turbines. Today, the salmon population in the northwestern region of the United States is approximately 13% of what it was at the end of the 1800s.

Conservationists are looking for ways to improve fish ladders so that more fish can make the trip home to spawn. Perhaps more flat "rest" areas along a ladder's steps would allow fish to make the leaps more gradually, and higher sides would protect hatchlings going downstream. Until a better way is found, fish ladders seem to be the only way salmon and other SPECIES can make it up and over the dams that stand between them and their spawning grounds. [*See also* HABITAT LOSS and HYDROELECTRIC POWER.]

Floodplain

▶A long, flat strip bordering a river that overflows its banks. A *flood* is any high stream flow that spills out of its channel. Most rivers

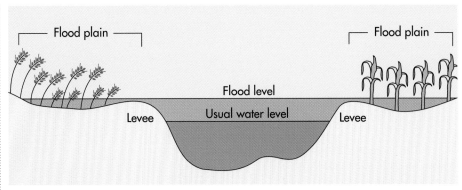

◆ When the river overflows its banks, water covers the floodplain, the land on both sides of the river.

overflow their banks every two or three years when the volume of water in the river increases rapidly for a short period. This increase may be caused by unusually rapid melting of a lot of snow in spring or by seasonally heavy rain. Rushing down hills, the flood waters pick up particles of rock and SOIL and carry them along. When the water reaches flatter ground, it slows and the particles fall to the bottom, where they are deposited as SEDIMENT.

Flooding moves sediment, made up mainly of clay and **silt**, from the surrounding hills into the river and onto the floodplain of a valley. The

land where the sediment is picked up by the water is said to be suffering from water EROSION. Flooding may deposit as much as 0.4 inches (1 centimeter) of sediment a

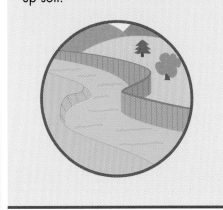

◆ Floodplains are areas of land covered by floodwater.

year on a floodplain. It also forms sandbars in the river itself, sometimes altering the course of the river flow.

BENEFITS PROVIDED BY FLOODPLAINS

Some floodplain soils may provide fertile ground for plant growth. A famous example is the long floodplain of the Nile River in Egypt. Many kinds of crops are grown in the floodplains. Floodplain WETLANDS also provide resting stops and feeding areas for a wide variety of migrating waterfowl.

PROBLEMS PRESENTED BY FLOODPLAINS

Channelization is the process of altering river banks in one of two ways. The first way is to construct dikes and embankments. The second way is to straighten, widen, or deepen the river to increase its ability to handle its water burden. The practice may have various goals, such as increasing the stream's capacity to hold floodwater or guiding the stream in particular directions. Some of the world's largest rivers are lined by huge embankments, like those that run about 600 miles (1,000 kilometers) beside the Nile; 870 miles (1,400 kilometers) along the Red River in Vietnam; and 2,800 miles (more than 4,500 kilometers) in the Mississippi Valley.

Channelization often achieves its goal, but it also has undesirable side effects and may create unexpected problems. Channelization reduces the number of habitats in the river and on its banks, the variety of organisms that live there, and

the water's ability to purify itself. Because water has to go somewhere, the dikes and **levees** that are built to prevent flooding may actually contribute to flooding upstream. In addition, when moving water spreads out over a floodplain, it slows. By forcing it back into a channel, an embankment speeds the water flow up again and delivers the water faster to areas downstream. These downstream locations may flood after channelization, even though they never did before.

Floodplains are natural safety valves, areas where a river has overflowed its banks during every flood for thousands of years. Building DAMS and levees does not always prevent floods. For instance, the Aswan High Dam in Egypt was built to hold back the floodwaters of the Nile and to provide ELECTRICITY. As a result, the floodplain of the Nile is no longer fertilized by the floods. Instead, the sediment-laden floodwater is stopped by the dam. As a result, the sediment falls to the bottom of Lake Nasser, behind the dam, which is fast filling up. Because of the sedimentation, Lake Nasser now holds much less water for drinking and IRRIGATION than its builders anticipated.

Possibly the worst problem with building dams and levees to hold back floodwater is that they may burst, destroying property and in some cases taking lives. Once a river is controlled, people tend to settle on the floodplain below. Thousands of people in India, Pakistan, and North America have been injured or killed in floods caused by bursting dams and levees. [*See also* SEDIMENTATION.]

Flowering Plant

▶A PLANT that reproduces sexually and develops seeds. Flowering plants are classified as *angiosperms*. This group of plants is the largest and most well-known group of plants on Earth today. Many types of plants are classified as angiosperms, including grasses, all fruit-bearing trees, and LEGUMES such as peas and peanuts. There are more than 230,000 SPECIES of angiosperms on Earth. They live in almost all types of HABITATS.

TYPES OF FLOWERING PLANTS

Flowering plants can generally be divided into two groups: monocotyledons (called *monocots*) and dicotyledons (called *dicots*). The differences between the plant types have to do with the number of seed leaves, or cotyledons, contained within the seed. For example, monocots contain only one seed leaf. This group of angiosperms contains about 60,000 species. Traits of monocots include leaves with parallel veins and flowers made up of three petals or multiples of three. Familiar monocots include grasses, lilies, orchids, and palm trees.

The seeds of dicots have two seed leaves. Dicot leaves show a netlike pattern of veins. Their flower parts are in multiples of four or five. There are approximately 170,000 known species of dicots. Familiar dicots include fruit plants, such as apple, plum, and peach trees; hardwood trees, such as oak and maple; vegetables, such as broccoli, cabbage, and lettuce; and wildflowers, such as dandelions, goldenrod, asters, and sunflowers.

THE ROLE OF FLOWERS

The diversity of flower forms is evidence of the great success of flowering plants. There are probably as many different shapes, sizes, colors, and configurations of flowers as there are species of flowering plants.

Despite the amazing diversity of flowers, all share a common function—reproduction. Most flowers have four parts: *petals, sepals, stamens,* and *pistils.* Some of these parts are involved in fertilization and seed production. Others are important for POLLINATION.

Petals are the leaflike, brightly colored structures that surround the flower. Their brilliant colors, odors, and shapes attract INSECTS, BIRDS, and bats, which all help pollinate the plant.

Sepals are also leaflike structures. They are usually green and circle the flower stem beneath the petals. Some sepals, such as those of tulips, may be colored like petals. Sepals form a cuplike structure called a *calyx,* which helps protect the developing flower bud from insect damage and dehydration.

Within the ring of petals and sepals are the reproductive structures of the flower. Stamens are long, thin structures that are the male reproductive parts of flowers. Most flowers contain several stamens. At one end is a thickened structure called the *anther.* The anther produces pollen, the male reproductive cells of angiosperms. When the developing pollen grains mature, the anther splits open to release them.

The female part of the flower is the pistil. This complex structure is made up of several parts. The *stigma,* at the top of the pistil, is a sticky surface on which pollen grains land and grow. The bottom portion of the pistil is a bulging, rounded structure called the *ovary.* Eventually, the ovary becomes the fruit. Connecting the stigma and ovary is a thin stalk called the *style.* Most flowers contain all four structures—petals, sepals, stamens, and pistils. These are known as complete flowers. Incomplete flowers have either stamens or pistils, but not both. An example of a plant with incomplete flowers is the cottonwood tree.

POLLINATION AND FERTILIZATION OF FLOWERS

Pollination is the transfer of pollen from the anther to the stigma. Flowers can be pollinated in different ways. In general, a flower's structure is directly related to how it is pollinated. For instance, wind-pollinated flowers are usually long and thin, with small petals or none at all. Wind-pollinated plants, such as corn, wheat, and ragweed, produce huge amounts of pollen that can be scattered across large areas.

Many species of angiosperms are pollinated by animals, such as bees, flies, butterflies, beetles, birds, and bats. These species produce flowers with much more elaborate, and sometimes spectacular, shapes. In many species, the structure is directly related to the type of animal that pollinates it. Pollination occurs when animals pick up sticky pollen on their bodies and distribute it to other flowers. It can also happen when an animal brushes against an anther, releasing a cloud of pollen into the air where it can land on the same flower or others close by.

The relationships between flowering plants and their animal pollinators, particularly insects, demonstrate COEVOLUTION, the process in which different organisms evolve ADAPTATIONS in response to each other. For example, flowers pollinated by hummingbirds are usually tubular and bright red or yellow—colors that attract hummingbirds. These types of flowers also tend to give off little scent because hummingbirds do not have a well-developed sense of smell. Flowers that are pollinated by bats, such as banana flowers, produce a lot of nectar, which attracts hungry bats.

Once a pollen grain reaches the stigma, several events take place before fertilization occurs. First, the pollen grain grows a long tube inside the pistil and into the ovary. Next, the sperm cells inside the pollen grain travel down the tube to reach the eggs within the ovary. Here fertilization occurs. In most plants, the pollen tube growth and fertilization occur fairly quickly. In barley plants, for instance, less than an hour elapses between pollination and fertilization. In corn plants, the process takes about 24 hours.

◆ Apple blossoms contain reproductive organs that produce seeds after pollination and fertilization occur.

◆ The coffee flower produces red berries, each of which contains two beans, or seeds.

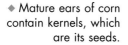

◆ Mature ears of corn contain kernels, which are its seeds.

After fertilization takes place, most of the flower parts begin to die and the seed develops. As the seed develops, the ovary surrounding the seed swells, beginning the process of fruit formation.

SEED DISPERSAL IN FLOWERING PLANTS

Much of the success of flowering plants in terrestrial ENVIRONMENTS is due to their sometimes elaborate mechanisms for dispersing seeds. Dispersal of seeds is important because it reduces COMPETITION for water, sunlight, and nutrients between new plants and their parent plants.

Fruits are one of the most common mechanisms for spreading seeds. For instance, when animals such as birds and squirrels eat fruits, they aid in the dispersal of seeds by tearing apart fleshy fruits or carrying fruits to new locations. When seeds are eaten, they often pass through an animal's digestive system and are later deposited in SOIL with the animal's droppings.

Flowering plants often depend on the wind for dispersing seeds. Wind-dispersed seeds, such as those from milkweed and dandelion plants, have parachutelike structures or wings that enable them to be carried by the wind. Other plants, such as cockleburs, produce seeds that are surrounded by a casing of tiny hooks, enabling them to hitch rides on animals and passing humans by clinging to fur or clothes. [*See also* AIR POLLUTION; AUTOTROPH; BIODIVERSITY; BIOGEO-CHEMICAL CYCLE; COMMENSALISM; EN-DANGERED SPECIES; and SYMBIOSIS.]

Fly Ash

D Very fine particulate matter suspended in the exhaust gases that are produced by fuel-burning power plants, municipal waste combusters, and incinerators. After FUELS or other materials are burned, some residue, or ash, remains. The heavier particles stay behind as *bottom ash*. The lighter particles—*fly ash*—may be carried into the air or collected in various types of AIR POLLUTION control devices.

In high concentrations, fly ash forms grit on buildings, clothes, and cars. For people who live near fly-ash-producing places, it can be an expensive and irritating type of air pollution. Fine particles of fly ash can penetrate the lungs, posing a health hazard. Heavy deposits may also affect PLANT growth. Most industrial fly ash is collected and dumped in waste sites. [*See also* FOSSIL FUELS; GARBAGE; HAZARDOUS WASTE; PARTICULATES; POLLUTION; and SCRUBBER.]

Food Chain

D A series of organisms in an ECOSYSTEM that are linked to each other through their needs for energy and nutrients. Food chains vary from one ecosystem to another. However, all food chains have three types of organisms in com-

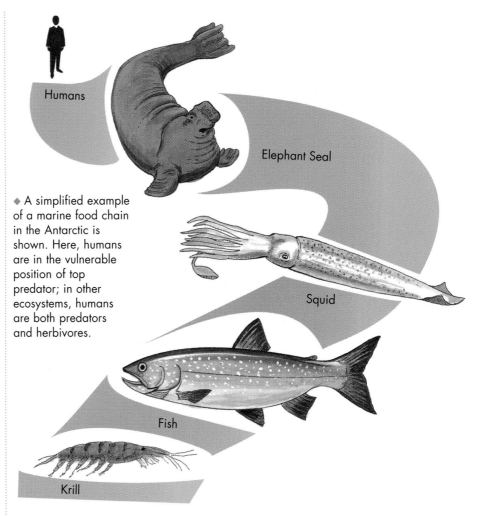

◆ A simplified example of a marine food chain in the Antarctic is shown. Here, humans are in the vulnerable position of top predator; in other ecosystems, humans are both predators and herbivores.

Humans

Elephant Seal

Squid

Fish

Krill

mon: PRODUCERS, CONSUMERS, and DECOMPOSERS.

Producers are organisms that obtain their energy through the process of either PHOTOSYNTHESIS or chemosynthesis. Such organisms include PLANTS, ALGAE, and protists (photosynthetic organisms) and some types of BACTERIA and chemosynthetic organisms living near ocean vents, or openings in Earth's crust. Consumers are organisms that obtain their energy and nutrients by eating, or "consuming," other organisms. Decomposers are organ-isms, such as bacteria and FUNGI, that obtain their energy and nutrients by feeding on the remains of dead organisms. Because of the way they obtain their energy, decomposers are always the last organisms in any food chain. Producers, consumers, and decomposers are at different TROPHIC LEVELS, or feeding levels, in a food chain.

A typical food chain begins with producers, such as plants, that obtain and store the energy from the sun. HERBIVORES, organisms that feed directly on plants (or on other producers), are the next organisms in a food chain. Herbivores are a type of consumer. Because they are at the first trophic level of a food chain, they are often called *first-order consumers*. Herbivores may also be the last level of certain food chains until they die and decomposers begin their feeding processes. However, the herbivore level of a food chain is usually followed by one or more CARNIVORES, or animals that feed on other animals.

All carnivores are consumers, but the carnivores in a food chain may differ in how they get their food. For example, a carnivore may be either a PREDATOR, which is an animal that hunts and kills other animals for food, or a *scavenger,* which is an animal that feeds on the remains (carrion) of animals that have died of natural causes or been killed by predators. Depending on where they feed in a food chain, carnivores (predators or scavengers) may be second-, third-, or fourth-order consumers.

Most food chains are short, involving no more than three to five different kinds of organisms at different trophic levels. The length of a food chain is largely determined by the amount of energy that can be passed from one feeding level to the next. For example, as organisms eat at higher trophic levels, less energy (fewer calories) is available to the organism than if it fed at lower levels. Predators at the highest levels of the food chain may therefore be extremely vulnerable to changes in food supply and environmental disturbances. One result of such limits is that plants and

other producers are fairly abundant whereas, as ecologist Paul A. Colinvaux has stated, "big fierce animals are rare." [*See also* AUTOTROPHS; ENERGY PYRAMID; and FOOD WEB.]

Food Web

▶A biological model that shows the overall pattern of feeding and energy relationships in BIOLOGICAL COMMUNITY. A FOOD CHAIN shows the feeding relationship for one series of organisms. For example, in the food chain *grass → grasshopper → bird,* you can see that some of the energy and nutrients in grass are taken in by a grasshopper eating the grass. In turn, the grasshopper is eaten by a BIRD that obtains its energy and nutrients from the grasshopper.

A food web is a series of interconnected food chains. Unlike a food chain, a food web shows the feeding relationships among several groups of organisms. In this way, a food web includes information about the known feeding relationships among several interacting SPECIES in a community.

Most species do not eat only one kind of food. Thus, in a food web, an individual species may feed at different TROPHIC LEVELS in several related food chains. A single species may be connected to a large number of other species as either a PREDATOR or prey at different trophic levels. For example, a small OMNIVORE such as a raccoon

may eat several different species of PLANTS and animals. When eating a plant, the raccoon is feeding at the first trophic level. However, when feeding on animals, such as FISH, the raccoon may be feeding at the second, third, or even fourth trophic level. In turn, the raccoon may be the prey of a variety of predators that also feed at different trophic levels.

GENERAL FEATURES

Of the many food webs that exist in nature, scientists have at least partly described about 200. From these accounts, some general features of food webs can be noted. First although many food chains may be shown in a food web, the food chains themselves are short. Usually, there are no more than five links from a plant through intermediate species (which are both predators and prey) to the top predators in the ECOSYSTEM. A second feature common to food webs is that there seems to be a fairly constant ratio of two to three prey species for every predator species.

In large food webs, each species may not necessarily have close relationships with every other species. This does not mean, however, that feeding relationships in a large food web are not important. In a food web of any size, the roles of some species, often called KEYSTONE SPECIES, may be critical for the entire community.

CONTROVERSIES

Food chains and food webs have been methodically studied only since the 1920s. Only a fraction of the world's ecological communities have been described. In most cases, these descriptions are simplified and partial because of the large numbers of species involved. For example, a DESERT food web may involve as many as 2,000 species. The food web for a tropical RAIN FOREST community probably involves even more organisms. Most scientific descriptions of food webs cannot trace such a large number of connections. As a result, even the general features of food webs are matters of debate.

Scientists also debate whether a complex, diverse food web is more stable than a simple food web that involves fewer species. In the simple food webs of the ARCTIC, population explosions and crashes are common in small animals such as lemmings, snowshoe hares, and some of their predators. In contrast, the species involved in the complex food webs of tropical rain forests seem to have more stable populations. This observation led many scientists to believe that BIODIVERSITY promotes stability.

However, in the last few decades, mathematical data, laboratory experiments, and field studies have shown that the relationship between biodiversity and stability may not be so simple. Such studies have shown that the stability of an ecosystem seems to depend more on the actual species involved than it does on the SPECIES DIVERSITY of an ecosystem.

HOLES IN THE WEB

CONSERVATION biologists and WILDLIFE managers must often predict how the loss or addition of a species will affect a biological community. However, it is not always clear how such changes affect the food web. In some cases, the NICHE of an

◆ A simplified food web for a forest and pond in the north temperate zone is shown above. Note that many food web interactions are in the pond.

dominant species type of organism that is especially abundant in one of the trophic levels of a food web. In the American prairies, for example, bison were once the dominant species of herbivores.

feeding guild a group of species in a food web that all feed on a certain resource in a similar way; for example, fruit-eating birds.

extinct species may be taken over by another species in the same **feeding guild**. For example, the coyote has replaced the wolf as the top predator in many FORESTS of the eastern United States.

Some scientists believe that even a **dominant species** may disappear from a community without causing large changes in the food web. Evidence of this was seen in the case of the American chestnut tree. Before 1910, the American chestnut made up about 40% of the canopy of the DECIDUOUS FORESTS of the eastern United States. The species then almost disappeared as

a result of a disease called *chestnut blight* (caused by FUNGI). When American chestnuts were in abundance, they served as the main food source of certain moth species. Most of these species survived by feeding on other plants.

In some cases, the loss or addition of a species to a community does affect the food web. For example, PESTICIDE use frequently kills its intended pests, as well as predators of the pests. With fewer predators, the original pest species may actually begin to thrive. Similarly, the EXTINCTION of a keystone species or the introduction of an EXOTIC SPECIES can cause great changes in the feeding relationships in a community. Thus, wise management decisions depend on thorough knowledge of the structure and function of food webs. [*See also* ADAPTATION; AGRICULTURAL POLLUTION; AGROECOLOGY; AUTOTROPH; BACTERIA; BIOLOGICAL CONTROL; CLIMAX COMMUNITY; COEVOLUTION; COMMENSALISM; DETRITUS; ENERGY PYRAMID; ENVIRONMENTAL EDUCATION; FAMINE; GRAZING; HABITAT LOSS; HUNTING; INTEGRATED PEST MANAGEMENT (IPM); MUTUALISM; NITROGEN CYCLE; PARASITISM; PEST CONTROL; PHYTOPLANKTON; PLANKTON; and SYMBIOSIS.]

Forest

▌A large land area on which trees are the dominant plant life. A forest also contains other PLANTS, like bushes, mosses, herbs, and wild-

flowers, along with animals, such as BIRDS, REPTILES, INSECTS, and MAMMALS. Most forests require 30 inches (75 centimeters) of rain yearly to support plant and animal life, so ANTARCTICA is the only continent that does not sustain forests.

Millions of years ago, forests covered more than 60% of Earth's land surface. Then Earth's CLIMATE changed, GLACIERS raked its surface, and humans cleared trees for firewood and to make way for farms and cities. Today about 30% of the land surface is covered with forests of different types, depending on their locations. For example, a hot and humid Central American RAIN FOREST is very different from a CONIFEROUS FOREST in northern Canada, where towering fir and spruce trees thrive in the cold, snowy climate.

All forests have plants, animals, and other types of organisms that interact with one another and their surroundings. The green plants use sunlight, CARBON DIOXIDE, and MINERALS from the SOIL to create their food through PHOTOSYNTHESIS. In this food-making process, the OXYGEN emitted as a waste product is the life-sustaining gas needed by all organisms, some of whom feed on the green plants. The plant eaters are, in turn, eaten by other animals. As the organisms of a forest die, nature recycles them within the ECOSYSTEM. Their remains decay, with the aid of BACTERIA and FUNGI, and enter the SOIL as MINERALS that are used by plants to use in making food.

Forests are areas of constant COMPETITION and cooperation. Plants compete for sunlight, water, and minerals from the soil; animals compete for food and shelter.

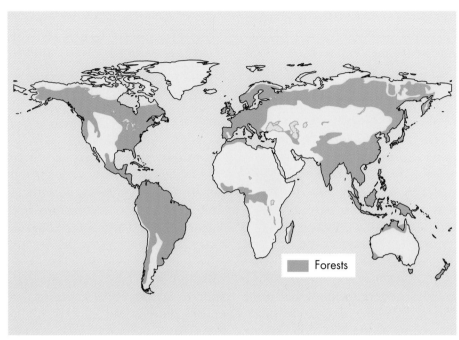

◆ Approximately 30% of Earth's land surface is covered with forests of different types.

Animals that are dependent on plants for food cooperate with the plants by spreading the plants' seeds. In total SYMBIOSIS, fungi feed on trees while helping them absorb water and nutrients. Although individual forest organisms die, the forest as a whole survives.

FOREST LAYERS

Most fully developed forests have five layers, created by plants of uneven heights: the canopy, understory, shrub layer, herb layer, and forest floor. The canopy contains the crowns—top branches and leaves—of the tallest trees, which receive the most sunlight. Mammals, insects, and birds that eat leaves or fruit make their HABITAT in the canopy.

Understory trees are shorter than those in the canopy. Many are SPECIES that thrive in shade; others are young trees that will someday become part of the canopy. Below the understory are shrubs—woody plants that branch out from a single stem. They never grow as tall as trees, but if the canopy is open enough to allow sunlight through, shrubs can grow thick and full to provide homes for birds and insects.

The herb layer has FERNS, mosses, wildflowers, and soft-stemmed plants. An open canopy allows them to grow thick and lush, supplying habitats for insects, turtles, mice, snakes, and ground-nesting birds. The forest floor is covered with moss, fallen leaves and twigs from upper layers, animal wastes, and dead animals. These dead materials, called DETRITUS, are used by forest DECOMPOSERS as food. Insects, worms, spiders, fungi, and

bacteria help break down this **debris** to release nutrients into the soil for new plant growth.

CLASSIFYING FORESTS

Forests are classified in several ways. One way is according to the leaf characteristics of dominant trees. For example, needle-leaf forests have mostly trees with long, thin, needlelike leaves. Broadleaf forests have mostly trees with wide, flat leaves. A second way to classify forests is according to how often the dominant species loses and grows leaves. For example, in a DECIDUOUS FOREST, dominant trees shed leaves once a year. In an evergreen forest, dominant trees shed and grow new leaves continually. A third classification is by dominant wood type: **hardwood** or SOFTWOOD. This method of classification is often used by people working in the FOREST PRODUCTS INDUSTRY.

Most scientists also classify forests into five or more groups according to climate and formation. The five most common groupings using these characteristics are: tropical rain forests, tropical deciduous forests, temperate deciduous forests, temperate evergreen forests, and the TAIGA.

Most tropical rain forests are near the equator, where the climate is warm and wet all year. These forests have a vast variety of trees—as many as 1,000 species within a 1-square-mile (2.6-square-kilometer) area. Most trees in a tropical rain forest are all broadleaf evergreens. The canopy is about 150 feet (46 meters) above the forest floor. The forest floor here is nearly bare because of limited sunlight and rapid decay of debris. Many scientists believe that rain forests house about 50% of all terrestrial plant and animal species on Earth. Animals include snakes, monkeys, birds, bats, insects, lizards, mice, and sloths.

Tropical deciduous forests develop in tropical and subtropical regions. They resemble rain forests and have similar WILDLIFE. However, unlike tropical rain forests, tropical deciduous forests have distinct wet and dry seasons. During the dry season, many trees shed their leaves. Slightly cooler than rain forests, these forests have thick layers of palm and bamboo trees below their 100-foot (30-meter) canopies.

Temperate deciduous forests, in which most trees are broadleafs, grow in areas with warm summers and cool winters. The herb layer has two periods of growth: plants that sprout in early spring, before trees are covered with leaves, die in summer, and are replaced by plants that thrive in shade. Forest animals include bears, deer, wolves, and hundreds of birds and small mammals, many of whom migrate or hibernate in winter.

Temperate evergreen forests are common in coastal areas with mild winters and heavy rainfall. Many are located along the Gulf and northwestern coasts of the United States. Snails, frogs, and other animals that live near water are numerous in these forests.

Taigas, also known as boreal forests, grow in areas with extremely cold winters and short, cool summers. Due to a short growing season, these forests have fewer layers. A 75-foot (23-meter) canopy of needle-leaf evergreens, like spruce, pine, or fir, grows above a sparse shrub layer. Moss and LICHENS grow in abundance on the forest floor. These organisms often make their home on bare rock, tree trunks, or among the debris that accumulates on the floor. Animals include beavers, porcupines,

◆ In some forests, such as this one in Maine, both coniferous and deciduous trees grow.

◆ Numerous Douglas fir trees may be found in this old-growth forest near Jefferson Creek, Washington.

rabbits, moose, caribou, foxes, wolves, bears, ducks, loons, owls, and woodpeckers. Some animals migrate in winter; others hibernate. Still others, such as the snowshoe hare, have adaptations that allow them to survive the cold, snowy winter conditions.

THE IMPORTANCE OF FORESTS

Prehistoric people hunted forest animals and gathered berries and nuts for food. They used tree branches for shelter; made clothing from animal skins and weapons from animal bones; and burned wood as fuel. Today, people still depend on forests to supply thousands of products, including lumber, toys, toothpicks, furniture, paper, paint, PLASTICS, tires, nuts, and clothing.

Forests also have enormous environmental value. The organic debris on a forest's floor absorbs huge amounts of rainwater, preventing the loss of SOIL by EROSION. The rainwater filters through the soil, often filling underground AQUIFERS that provide clean, fresh water for wells, streams, and lakes. Forests influence Earth's climate by reducing wind force over land. In addition, the cooling effect above a forest can increase rainfall by about 3% over that of open land.

One of the greatest environmental rules of forests is their contribution to the CHEMICAL CYCLES that sustain life and regulate WEATHER and climate. For example, trees renew the air of the atmosphere by giving off oxygen and removing carbon dioxide, one of the GREENHOUSE GASES. A buildup of greenhouse gases in the atmosphere could cause substantial changes in Earth's climate. Because of their economic and environmental importance, it is essential that people protect and preserve forests through CONSERVATION, proper management and replanting. [*See also* ADAPTATION; AGROFORESTRY; BIODIVERSITY; BIOGEOCHEMICAL CYCLE; BIOME; CLEAR-CUTTING; CLIMAX COMMUNITY; DECOMPOSITION; DEFORESTATION; DESERTIFICATION; EXTINCTION; FIRE ECOLOGY; FORESTRY; FUEL WOOD; GLOBAL WARMING; GREEN REVOLUTION; GRIZZLY BEAR; HABITAT LOSS; HUNTER-GATHERER SOCIETY; MIGRATION; MUIR, JOHN; NATIONAL PARKS; NORTHERN SPOTTED OWL; OLD-GROWTH FOREST; PRECIPITATION; PRESCRIBED BURN; PUBLIC LAND; SUSTAINABLE DEVELOPMENT; and TREE FARMING.]

Forest Fire

◗ A fire that burns surface vegetation, including trees, shrubs, GRAZING land, flowers, weeds, and underbrush in a FOREST. Lightning is the only natural cause of forest fires, but it is responsible for less than 10% of all forest fires that occur in the United States. The rest, which are caused by humans, can generally be prevented.

The majority of forest fires are caused by people who drop lighted cigarettes or matches. Other forest fires are caused by machines such as AUTOMOBILES, trains, or logging equipment that overheat or spark near dry vegetation. For example, fires sometimes begin when a car's

◆ Rangers of the U.S. Forest Service help fight forest fires.

hot tailpipe contacts drought-dry grass. Fire can also start from campfires that are not set up safely or extinguished after use. Another cause of forest fires is *arson,* fires that are deliberately set.

Once woody or dry plant material begins to burn, the heat spreads to nearby combustible matter that ignites, spreading the fire farther. Only favorable WEATHER conditions or the firefighting techniques of humans can stop the fire from destroying everything in its path.

From watchtowers or airplanes, forest rangers are always on the lookout for fires. If a fire is spotted, its type and location is radioed to fire fighters. There are several types of forest fires. Surface fires, the most common, burn through small PLANTS and debris on the forest floor. A ground fire burns the organic matter under the debris on the forest floor. A crown fire consumes mainly the crowns of trees and shrubs.

Putting out a forest fire requires removing potential FUEL from its path. Water or a chemical is sprayed on the fire at ground level or dropped from the air by helicopter or plane. The liquid slows the blaze and cools it so that fire fighters can get close enough to dig a **fireline** across it or in front of it. Using axes, shovels, and bulldozers, fire fighters clear vegetation and scrape away SOIL around the fire. In hard-to-reach areas, "smoke jumpers" parachute in to dig the fireline.

To prevent a fire from jumping the fireline, a **backfire** may be set that can burn the area between the fireline and the fire itself. To make sure smoldering embers do not reignite, all burnable material is cleared from the edges of the burned area.

FOREST FIRES AND THE ENVIRONMENT

Fire has both positive and negative effects on forests. It is nature's way of cleaning out debris so that more sunlight can reach the ground where young trees grow. Some pine trees, such as jack pines, need the heat of the fire to discharge their seeds. To achieve these positive results, foresters sometimes carry out a PRESCRIBED BURN, or controlled burning. If weather conditions permit, a small fire is set in the forest debris. The fire is monitored every second as it cleans out surface clutter, restricting the possibility of a major forest fire, and killing diseased plants, insect pests, and unwanted seedlings.

Fire can harm a forest because it leaves soil exposed to EROSION. Burning trees cannot emit OXYGEN; they give off CARBON DIOXIDE, one of the GREENHOUSE GASES scientists believe are responsible for GLOBAL WARMING. WILDLIFE in the ECOSYSTEM suffers HABITAT LOSS and either leaves the area or dies. However, out of disaster some good may come. It will be years before new trees reach the heights of old ones, but rebuilding begins with new plants and animals that return to feed on the plants that may develop. [*See also* FIRE ECOLOGY; FOREST; GREENHOUSE EFFECT; NATURAL DISASTERS; and SUCCESSION.]

Forest Products Industry

▶Manufacturing activities that use a variety of products from trees, including lumber, lacquer, paper, and even Ping-Pong balls. In the United States, more than 1.5 million people work in 50,000 manufacturing plants that produce about $120 billion worth of forest products each year. To supply the materials needed to make their products, manufacturers own about 70 million acres (28 million hectares) of forestland in the United States. They also have government leases that allow them to harvest trees from state and NATIONAL FORESTS and buy logs from private owners of small wooded areas.

Most forest products are made from the wood of trees. Some are from the bark, gum, fruit, seeds, leaves, and sap. For example, bark from the cork tree is used to make bulletin boards, insulation, and bottle corks. Bark from other trees contains tannic acid, which is used

THE LANGUAGE OF THE ENVIRONMENT

backfire a fire deliberately set to stop an advancing fire.

fireline a wide strip around a fire that is cleared of brush, logs, and trees. Firelines are dug to slow the progress of a fire.

in curing animal hides for leather products. Some bark is made into FUELS or mulch—ground covering that retains water around PLANTS, adds nutrients to SOIL, and reduces the growth of weeds.

Some seeds and fruits of trees, such as the almond, walnut, and pecan, are edible. Seedpods from the kapok, or silkcotton tree, supply material that is used for stuffing sleeping bags and jackets. Milky latex that is drained from trees supplies natural rubber for making tires and balloons. Maple sap becomes a sweet syrup. Oils from evergreen and eucalyptus leaves are used to make perfumes, soaps, and cleaners. Parts of other trees are used in making medicines.

FOREST PRODUCTS THROUGH THE YEARS

For thousands of years, forest products have fulfilled many human needs. Prehistoric people ate berries and nuts, built shelters using tree branches, made clothing from plant material, and began using wood as fuel around 500,000 B.C. Today, forests provide the raw materials that are used to make thousands of things we classify into four major groups: wood products, fiber products, chemical products, and fuel products.

Wood Products

Wood products are made from solid wood or pieces that are large enough to be recognizable as wood. They include posts; pilings; lumber; railroad ties; sporting goods, such as baseball bats and hockey sticks; chopsticks; toothpicks; matches; furniture; and tongue depressors.

Fiber Products

Fiber is the slender threadlike material in wood. Wood chips are reduced to a fibrous pulp to make paper and paper products such as books, bags, cartons, and tissues. Wood fiber is also used to make plywood insulation, mulch, dietary fiber for cereal, furniture, siding, and paneling. Still other fibers, such as those from kapok, are used to make articles of clothing or insulating materials.

Chemical Products

Chemical wood products are made from cellulose—the primary material in the wood fibers—and lignin—the material that holds the fibers together. From cellulose, manufacturers create paints, lacquer, PLASTICS, piano keys, and the rayon that is used to make clothes, tires, cellophane, and photo film. From lignin, they make inks, dyes, and artificial vanilla.

Fuel Products

FUEL WOOD is still the main energy source in some countries. It is an ALTERNATIVE ENERGY SOURCE in others. Fuel products include split and dried logs, charcoal, and compressed wood pellets. The thick liquid residue produced from making wood pulp can also be burned as fuel.

INDUSTRY AND THE ENVIRONMENT

Only sustained yield, a balance between the yearly harvesting of trees and the growth of trees that will eventually replace them, can guarantee a continuous supply of wood and other forest products from Earth's forests. Current timber harvesting practices by some forest products companies endanger the future wood supply and threaten our ENVIRONMENT. DEFORESTATION and CLEAR-CUTTING, which create areas

◆ Many products we use daily are made from raw forest materials.

where burned or cut-down trees emit CARBON DIOXIDE as they decay, add to GREENHOUSE GASES in our ATMOSPHERE. At the same time, this practice may lessen the amount of OXYGEN added to the air because the trees are no longer carrying out PHOTOSYNTHESIS. Some scientists believe that a buildup of greenhouse gases, such as carbon dioxide, will lead to a GLOBAL WARMING and other changes in the environment of our planet. [*See also* AGRO-FORESTRY; CONIFEROUS FOREST; DECIDUOUS FOREST; ECOTOURISM; FOREST; FORESTRY; GLOBAL 2000 REPORT, THE; GREEN POLITICS; GREEN REVOLUTION; OLD-GROWTH FOREST; RAIN FOREST; and SILVICULTURE.]

Forestry

▌The cultivation and management of FORESTS for human use. Human beings use tremendous amounts of wood and wood products such as paper. Firewood is the major source of fuel for an estimated 1.5 billion people in developing countries. Worldwide, timber is harvested at a rate of 110 cubic inches (1800 cubic centimeters) per person per day. A disproportionate amount of this harvested wood—3.5 times the per-capita average for the rest of the world—is used by people in the United States. The purpose of forestry, in the face of such enormous demand, is to ensure a continuous supply of lumber, wood **pulp**, and other forest products.

◆ The great demand for forest products has often resulted in the harvesting of trees in old-growth forests.

Forestry includes SILVICULTURE, which is the science of the growth and management of trees. The specific aims of forestry are to prevent and control FOREST FIRES, to protect forests from insect pests and disease, and to prevent waste in lumbering. Important forestry practices are reforestation, the replanting of burned or cutover lands, and afforestation, the planting of new forest.

In the final analysis, however, forestry is forest management, and forest management comes down to the harvesting of trees. Many different tree-harvesting strategies are

used, but two especially common methods are CLEAR-CUTTING and selective cutting. In clear-cutting, all trees are removed from an area. The great advantage of clear-cutting is purely financial: because it is indiscriminate, relatively fast, and requires little in the way of planning, skill, or access roads, it is the least expensive of all harvesting methods, increasing yield per acre and shortening the time required to reforest an area with desirable (generally fast-growing) types of trees. Clear-cutting also fragments and devastates wildlife HABITATS and the subsurface environments in which FUNGI, worms, and microorganisms condition SOIL; leaves areas vulnerable to wind and water EROSION, sediment water pollution, and flooding; and destroys landscapes.

In selective cutting, middle-aged or mature trees are harvested. These are removed individually or in small groups. The gaps created are no wider than the height of the remaining trees, which regenerate the harvested patches. Selective cutting reduces crowding, allows younger trees to grow, and maintains a stand of trees of different species and uneven ages. Done properly, selective cutting in itself is the least destructive method of harvesting, enabling forest land to provide one tree crop after another on a continual-yield basis. Unfortunately, more than any other method of tree harvesting, selective cutting requires the building of access roads, which are the main cause of soil erosion associated with logging operations.

Some lumber, paper, and other wood-using companies in this

THE LANGUAGE OF THE ENVIRONMENT

shelterwood cutting a method of forestry in which all mature trees are removed in cuttings of two or three over a period of 10 years. The first cut removes most mature canopy trees; unwanted tree species; and diseased, defective, and dying trees. This opens the forest floor to light and leaves enough mature trees to cast seed and to shelter growing seedlings. Some years later, after enough seedlings take hold, more canopy trees are removed, leaving some of the best mature trees to shelter young trees. When the young trees are well established years later, the remaining mature trees are cut, and the even-aged strands of young trees are allowed to grow into maturity.

pulp a mixture of ground-up, moistened material.

◆ Cottonwood trees are being logged in Missouri. The logs are stacked before they are loaded into trucks.

country grow and harvest crops of trees on their own lands, which are certified under the American Tree Farm System, an industry-sponsored program with the stated purpose of recognizing good forest management. But private individuals and companies are also permitted to lease tracts of NATIONAL FORESTS to harvest timber.

The United States first adopted a national forest-conservation policy in 1891, when the president was authorized to set aside certain forest areas, which are now known as national forests. Responsibility for these PUBLIC LANDS was vested in the Division of Forestry under the DEPARTMENT OF AGRICULTURE. The Division of Forestry later became the FOREST SERVICE.

In theory, the national forests are carefully managed to produce the greatest possible benefit for the greatest number of people. Management of the forests has been mandated by Congress for MULTIPLE USE, a term meaning a combination of extraction of resources, recreation, and protection of WATERSHEDS and WILDLIFE. Critics of the multiple-use policy charge that, in fact, it does little more than justify wasteful and expensive exploitation of public lands by the timber industry and other private businesses.

In recent years, industry and government foresters have begun advocating "New Forestry" or "New Perspectives," which emphasizes an ECOSYSTEM approach to forest management, emphasizing concern for the ecological health and diversity of forests over the production of maximum harvests of logs. Specifically, the movement recommends **shelterwood cutting** over clear-cutting; less frequent as well as more selective harvests—at 350-year intervals rather than the standard 60 to 80 years—allowing dead logs and other organic debris in forests to decompose and replenish soil; and reducing erosion and protecting fish habitats by leaving wider buffer zones along streams. New Forestry seeks to extend protection of broad landscapes—up to 1 million acres (400,000 hectares)— across private and public boundaries. This is possible, of course, only if private landowners are involved in management decisions, and many individuals and companies vehemently oppose any outside interference in activities on privately owned land. [*See also* AGROFORESTRY; BIODIVERSITY; FOREST PRODUCTS INDUSTRY; OLD-GROWTH FOREST; PRESCRIBED BURN; and TREE FARMING.]

Forest Service

▮An agency of the U.S. DEPART-MENT OF AGRICULTURE (USDA), whose primary purpose is the CONSERVA-TION and protection of America's NATIONAL FORESTS and GRASSLANDS, or PRAIRIES. Originally named the Bureau of Forestry, the agency was set up under the direction of the U.S. DEPARTMENT OF THE INTERIOR with the responsibility for protecting forest reserves of the United States. The Bureau of Forestry was officially renamed the U.S. Forest Service in 1903. In 1905 it was made part of the USDA.

Gifford PINCHOT served as the first director of the U.S. Forest Service. He established the goals of the Forest Service, the promotion of the best use of forestland and the conservation of the NATURAL RESOURCES under its jurisdiction. In 1907, the U.S. forest reserves were formally renamed "national forests."

Today, the U.S. Forest Service manages more than 150 national forests and 20 national grasslands. These areas cover about 190 million acres (76 million hectares) of land. Management of these lands includes the protection of WILDLIFE and the supervision of any GRAZING within the forests and on the grasslands, as well as preservation of the areas for use by future generations.

The U.S. Forest Service assists state and local governments and private landowners with their forestland and WATERSHEDS, or river channels. These areas cover another 480 million acres (192 million

hectares) of land. Rangers distribute PLANTS, help fight fires, and counsel the nonfederal forest owners on ways to reduce insect pests and diseases that could harm their trees. Service representatives also furnish information about harvesting and the best ways to sell forest products.

Between 1933 and 1943, the Forest Service ran a program called the Civilian Conservation Camps. This program allowed out-of-work Americans to earn a living while helping to preserve our natural resources. Since 1964, the Forest Service's Job Corps has offered the same kind of opportunity.

MANAGING FORESTS AND GRASSLANDS

Our national forests furnish about 25% of the wood harvested annually in the United States. This wood is used for lumber, construction, and paper, and for the products of other important industries. Forests are RENEWABLE RESOURCES. They can continually supply huge amounts of

◆ The U.S. Forest Service oversees grazing in public lands such as this one at Yellowstone National Park in Wyoming.

wood and other products and still not be used up, as long as the annual growth of trees is greater than the amount of trees cut or destroyed by disease, fire, and INSECTS. This plan of forest management is known as *sustained yield*.

To keep our forests vital and dynamic places with sustained yield, the Forest Service removes older trees and fosters new growth by tending seeds competing for space on the forest floor or by replanting seedlings of fast-growing trees. Crowded forest areas, where tree growth has become stunted, are thinned out to allow the remaining trees to grow faster and straighter as they compete for sunlight and other resources.

Much of the water supply in the United States flows through national forests and grasslands, which the Forest Service is responsible for protecting. Rangers plant ground cover on slopes around watersheds. They also keep plant growth trimmed to direct the flow of water and reduce chances for flooding and EROSION.

The Forest Service also issues permits to allow private farmers and ranchers to graze their animals on the nation's property. Rangers are responsible for monitoring such activity to protect against OVERGRAZING and to ensure that long-range plans for the forest, and grasslands are met. Rangers also maintain a system of lookout stations, trails, roads, and telephone lines on the government's land, control citizens' use of the land for picnics and camping, and help to rescue anyone who becomes lost or injured while in the national forests or on the national grasslands.

FOREST SERVICE RESEARCH

Besides the day-to-day activities performed by Forest Service employees, the agency is constantly seeking better ways to manage the national forests. For that purpose, ongoing research is conducted at eight regional experimental stations and at other specific project sites throughout the United States.

Research covers areas such as the propagation, care, and development of various trees and the types of insects and diseases that attack trees. At the Forest Products Laboratory in Madison, Wisconsin, Forest Service researchers are also exploring new ways to use forest products. [*See also* AGROFORESTRY; ECOSYSTEM; GRASSLAND; OLD-GROWTH PUBLIC LAND; and WILDERNESS.]

Fossey, Dian (1932–1985)

▌A primatologist who conducted extensive studies on the mountain gorillas of Rwanda and Zaire in Africa. A primatologist is a scientist who studies **primates**. Like other primatologists—such as Jane Goodall, who studies chimpanzees, and Biruté Galdikas, who studies orangutans—Dian Fossey engaged in her work through **field studies**. Also like Goodall and Galdikas, Fossey concentrated her studies on only one type of primate—the mountain GORILLA.

The mountain gorilla is an ENDANGERED SPECIES. Fossey began her studies of mountain gorillas in Zaire in 1967. At this time, the mountain gorilla was believed to be in danger of EXTINCTION. Estimates of mountain gorilla populations suggested that only about 500 individuals existed in the wild. As Fossey discovered, the dwindling numbers of mountain gorillas were caused largely by two factors—HABITAT LOSS and POACHING.

The loss of habitat of the gorillas resulted primarily from the intrusion of people into their territory. Many of these people were clearing forestland for use as farmland or for GRAZING animals. In addition, wood gathered from the areas was used as a FUEL source by the native people of the area.

The second threat to the gorilla population resulted from poaching. Many animals were killed for the purpose of obtaining their hands, which were considered collectors' items for use as ashtrays by many wealthy people throughout the world. In addition, some gorillas, especially infants, were captured for sale to ZOOS and private collectors located throughout the world. In both cases, for each gorilla that was captured or hunted, many others died. The deaths of several gorillas for the capture of only one occurred because gorillas, as Fossey observed, live in family units in which the gorillas work together to protect and care for each other and their young.

Through her studies of gorilla families, Fossey learned much about the needs of mountain gorillas and the social structure of these families. She wrote of her findings in her book, *Gorillas in the Mist*, which was published in 1983.

Through this book and her writings in several magazines, Dian Fossey brought the problems threatening the survival of mountain gorillas to the attention of people worldwide. This attention helped to reduce poaching activities by making gorilla parts less desirable. It also led many zoos to stop taking gorillas from the wild and to focus their attention on the breeding of gorillas already in captivity. Another result of Fossey's work was the development of mountain gorilla ECO-TOURISM, a creative strategy for both CONSERVATION and economic development in the areas in which mountain gorillas live.

Fossey spent much of her time within gorilla territory seeking out and destroying the traps set by poachers. She was also involved in the capture of several poachers, whom she turned over to government officials for prosecution and punishment. These activities made Fossey an enemy of poachers. On Christmas Eve of 1985, Fossey was discovered slain in her camp. Although many people believe her death was caused by poachers, a tracker in Fossey's employment and an American scientist working at

Fossey's camp were arrested for the crime. The tracker committed suicide while in prison. The American scientist was permitted to leave for the United States with the aid of U.S. officials. He was later convicted of the crime in absentia. [*See also* ENDANGERED SPECIES ACT; FUEL WOOD; HUNTING; INTERNATIONAL UNION FOR THE CONSERVATION OF NATURE (IUCN); and NATIONAL WILDLIFE REFUGE.]

Fossil Fuels

Energy sources—COAL, oil, and NATURAL GAS—resulting from the **fossilization** and **carbonization** of organisms. Most scientists believe that fossil fuels began to form millions of years ago, when the remains of PLANTS and animals were buried in SEDIMENT at the bottom of the seas that covered much of Earth's present-day land surfaces. Over time, heat and pressure compressed the sediments into sedimentary rock and cooked the organic matter, forming deposits of HYDROCARBONS. A few scientists think that fossil fuels formed when METHANE seeped up from deep within Earth and mixed with the plant and animal remains.

When burned, fossil fuels release the sun's energy stored long ago in the plants and animals. More than three-fourths of the world's energy is supplied by fossil fuels, but known supplies are dwindling. Fossil fuels are NONRENEWABLE RESOURCES; they cannot be replaced, unless we wait hundreds of millions of years for them to form.

COAL

Coal, the most abundant fossil fuel, forms from the remains of fossil plants. Deposits are normally in seams from 3 to 6 feet (0.9 to 1.8 meters) thick, but one deposit found in Wyoming has an average thickness of 100 feet (30 meters). At its thickest, the seam is 220 feet (67 meters) thick.

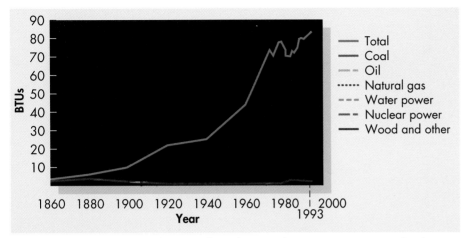

♦ Energy use has increased dramatically in the United States since 1860, with oil, natural gas, and coal being the fossil fuels most in demand. British thermal unit (BTU) is the quantity of heat required to raise one pound of water 1° F at a specified temperature.

Coal is ranked according to its stage of development. Each stage is marked by a further reduction in the percentage of moisture and an increase in the percentage of CAR-BON. The first stage is PEAT—brown, partially decayed organic matter in which twigs and other plant parts are still visible. Peat is almost 90% water, but it can be cut from the SOIL, shaped into small bricks or briquettes, dried, and burned. It has been used as a fuel in Germany, England, and Ireland for hundreds of years.

Under the right conditions, peat can develop into a very soft brownish-black coal called *lignite*, which contains about 50% water. The next grade, bituminous coal, is the most plentiful kind of coal. It is used not only for heating homes and businesses but also for making gas, a fuel called coke, and various chemical by-products of coal. The highest coal grade, anthracite coal, is a hard coal with the highest carbon content. Anthracite burns with a clean, almost smokeless flame. It takes three times as much lignite as anthracite to generate the same amount of heat.

In the United States, coal is generally burned in furnaces and boilers for heating purposes and to provide steam for generating ELECTRICITY or powering manufacturing machinery. Environmentalists warn that accumulated CARBON DIOXIDE released into the air by burning coal may result in a worldwide CLIMATE CHANGE. Ecologists are also concerned about land damaged by STRIP MINING, a practice in which coal is dug by gouging away the land around it.

OIL

Crude PETROLEUM, known as oil, developed from the remains of plants and animals compressed into sedimentary rock, such as sandstone and limestone. Through chemical processes, the rock formed a waxy material called *kerogen*, which separated into a liquid

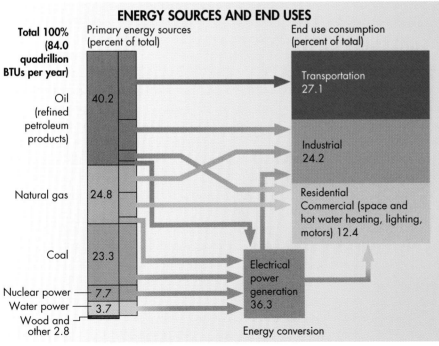

♦ The end uses of various types of fuel are shown above, together with an indication of how electricity fits into the overall pattern.

(oil) and a gas (natural gas) when underground temperatures rose higher than 400° F (204° C). Over time, the oil and gas moved upward through natural openings in the sedimentary rocks. Eventually the oil became trapped by a layer of nonporous rock called *caprock*. Unable to go farther, the oil condensed to form pools or reservoirs, and the lighter gas collected above the oil.

Some oil oozes to Earth's surface, but most must be pumped from the ground. Wells are drilled to tap into the oil reservoirs beneath the surface. The oil is under tremendous pressure and shoots to the surface. Although oil will burn as it comes from the well, most is sent to refineries to be processed into products such as gasoline, kerosene, tar, asphalt, and lubricating oils. In the United States, most oil is used for transportation, heating homes and businesses, and generating electricity.

Fires, spills, and pipeline leaks are environmental hazards that go along with drilling and transporting oil. Such hazards threaten the lives of millions of plants and animals.

NATURAL GAS

Natural gas, which is 83% methane, is normally found trapped close to crude oil, but some gas wells do not produce oil. Of all fossil fuels, natural gas is probably the most desirable. It can be sent by pipeline directly from wells to consumers, so it needs no storage facilities. It mixes easily with air, it is clean burning and less polluting than either coal or oil, and it generates

no ash. In the United States, natural gas is most often used for heating residential and commercial buildings, for transportation, for generating electricity, and for cooking. [*See also* ALASKA PIPELINE; AIR POLLUTION; ENERGY EFFICIENCY; EXXON VALDEZ; GLOBAL WARMING; and OIL SPILLS.]

Freon

▐▶Chemical compound used as a propellant in AEROSOL products and as a refrigerant. Freons are fluorine compounds that belong to groups of compounds called *fluorocarbons* and *chlorofluoromethanes*. These compounds are closely related to chlorofluorocarbons (CFCs). At room temperature, Freons exist as odorless and colorless liquids. In liquid form, they are useful as coolants in refrigerators and air-conditioning systems. When placed under great pressure, Freons exist as gases. In gas form, they are useful as propellants in products such as spray paints that are sold in aerosol form.

Around 1974, scientists began to suspect that CFCs and Freons were responsible for an expanding hole in the OZONE LAYER of the ATMOSPHERE. The ozone layer protects Earth and its organisms from much of the ULTRAVIOLET RADIATION given off by the sun. Many nations agreed to ban the production of CFCs by 1999.

Scientists are now using other chemical compounds in products that once contained CFCs and

Freons. Two groups of compounds now used are hydrofluorocarbons (HFCs) and hydrochlorofluorocarbons (HCFCs). These compounds may also damage the ozone layer, but not as much as CFCs and Freons. [*See also* OZONE and OZONE HOLE.]

Frontier Ethic

▐▶The belief that Earth's resources are meant for human consumption and thus can be used as if unlimited and without regard to how their use affects the ENVIRONMENT. There are two opposing schools of thought about the role of people in the natural world—the frontier ethic and SUSTAINABLE DEVELOPMENT.

An *ethic* is a set of standards used to help a person decide what is right and what is wrong. People who believe in the frontier ethic claim that economic growth is more important than CONSERVATION of nature. They believe that NATURAL RESOURCES are meant to be used, not conserved, and that new materials can be developed to replace resources that are used up. They also believe that people can use natural resources without concerning themselves with environmental problems such as POLLUTION. If air, land, and water do become polluted, they believe that people can develop technologies to control the pollution. In general, the frontier ethic is the belief that humans are separate from nature and that human success is measured by how much people can control nature.

FRONTIER ETHIC AND SUSTAINABLE DEVELOPMENT

People who believe in sustainable development believe that Earth has a limited amount of space for living things and resources. They believe that humans are a part of nature and do not have a right to control nature. The sustainable development view stresses that human success is measured by how well people live in harmony with the natural world.

The sustainable development ethic grew from a hypothesis proposed by British economist Thomas Malthus in 1789. Malthus believed that the human population will eventually grow too large for Earth to manage and sustain. When the population reaches this maximum size, or CARRYING CAPACITY, food supplies will be used up. Eventually, the population will be reduced in size through starvation, disease, and war.

The debate between the two opposing views—frontier ethic and sustainable development—has been going on for decades. Which view is correct is a matter of opinion. Because both of these views discuss how people operate in the natural world, they will certainly play an important role in determining the quality of human life in the future. [*See also* AGROECOLOGY; AGRO-FORESTRY; ENVIRONMENTAL ETHICS; GREEN POLITICS; INDUSTRIAL REVOLUTION; NATURAL RESOURCES; NONRENEWABLE RESOURCES POPULATION GROWTH; and SUSTAINABLE AGRICULTURE.]

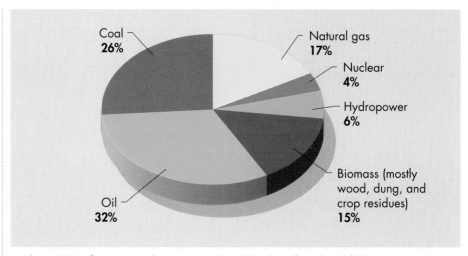

◆ About 75% of commercial energy used worldwide is from fossil fuels.

Fuel

❏Material that can produce energy for such uses as cooking, ELECTRICITY, and heat. Some fuels are found in nature; others are synthetic. Some are considered NONRENEWABLE RESOURCES, which means they cannot be replaced when they run out; others are RENEWABLE RESOURCES.

FOSSIL FUELS

COAL, oil, and NATURAL GAS are nonrenewable fuels formed over millions of years. These fuels formed from fossilized plants and animals that sank to the ocean floor and were covered by SEDIMENT. Intense pressure and high underground temperatures chemically changed the organic matter into CARBON and HYDROCARBONS.

Coal is mined from Earth's crust. Some mines are underground and entered by means of shafts; some are surface or strip mines, dug from the sides of hills. Coal, burned to produce heat or make steam for rotating **turbines** that generate electricity, affects the ENVIRONMENT in several ways. CARBON DIOXIDE from burning coal pollutes the air; STRIP MINING destroys large portions of land.

Crude oil (also called PETROLEUM) is trapped in absorbent sandstone or limestone. Wells are drilled through SOIL and rock to pump the oil up to the surface for use. Oil can be burned to provide heat as it comes from the well. However, most oil is sent to refineries and converted into products like gasoline, kerosene, diesel oil, asphalt, and lubricating oils. Burning oil poses the same AIR POLLUTION problems as burning coal. Oil fires and spills, and leaks from oil pipelines pose threats to the environment and to WILDLIFE.

Natural gas, usually trapped in rock with oil, is also brought to the surface by means of wells. Usually, natural gas is channeled through

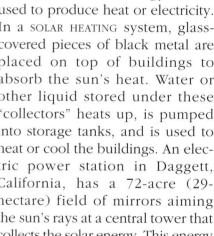

◆ Water and wind are both renewable resources that are harnessed to provide electricity.

pipelines directly to consumers. Gas, which is mostly METHANE, is clean burning and does not pollute the environment as much as coal or oil.

BIOMASS ENERGY

Wood, the oldest known fuel, is BIO-MASS—replaceable organic matter used to produce energy. Still important for cooking and heating in developing nations, wood is not a major fuel source in the United States. However, some paper factories burn sawdust, bark, and other wood waste to supply the power for their manufacturing processes. The PARTICULATES and gases given off by wood smoke can add to air pollution.

Other BIOMASS, like nutshells, rice, oat hulls, and other food pro-cessing by-products, is often burned to power plant machinery. In some cities, burning GARBAGE heats water to create steam for generating electricity. The burned garbage residue is a fine ash that is recycled into fertilizer, cinder blocks, and roadbed material.

NUCLEAR ENERGY

Nuclear fuels release energy by NUCLEAR FISSION or NUCLEAR FUSION. In fission, one atom splits into two smaller ones; in fusion, two small atoms join to create a larger one. In both processes, the formation of new atoms releases tremendous amounts of heat that can be used to create electricity. The constant threat of RADIATION from NUCLEAR POWER plants is an environmental concern.

SOLAR ENERGY

Sunlight that strikes Earth's surface can be used to provide free, clean, nonpolluting energy that will not run out. Such energy use is called SOLAR ENERGY. Solar energy can be used to produce heat or electricity. In a SOLAR HEATING system, glass-covered pieces of black metal are placed on top of buildings to absorb the sun's heat. Water or other liquid stored under these "collectors" heats up, is pumped into storage tanks, and is used to heat or cool the buildings. An electric power station in Daggett, California, has a 72-acre (29-hectare) field of mirrors aiming the sun's rays at a central tower that collects the solar energy. This energy is used to produce electricity.

WIND

Windmills have provided power for centuries. To use WIND POWER, wind turns a giant propeller attached to the shaft of an electric generator that feeds current into storage batteries. When the wind stops, the batteries supply needed power. The largest known windmill, with a 300-foot (90-meter) propeller, creates electricity for a power plant in Goldendale, Washington.

WATER

One of the oldest energy producers is flowing or falling water. It rotates old-fashioned waterwheels and modern turbines at DAMS to create HYDROELECTRIC POWER. However, many potential energy-producing waterfalls are in remote wilderness areas, making them very difficult to develop. Environmentalists believe such power projects may threaten NATURAL RESOURCES in an ECOSYSTEM located near hydroelectric power plants.

TIDAL ENERGY is used on the Rance River in Brittany, France. In this system, a dam pumps water into the ESTUARY at high tide. When enough water has built up, it is allowed to flow back through the dam's turbines to produce electricity.

GEOTHERMAL ENERGY

The energy in heat is called *thermal energy*. Much heat energy is stored in melted rock located beneath Earth's surface. Thus, heat energy contained in Earth is called GEOTHERMAL ENERGY. Earth's heat causes earthquakes and volcanic eruptions, and it can also be harnessed for use as fuel. For example, in Iceland, rising steam from hot springs is used to provide low-cost power to heat homes.

SYNTHETIC FUELS

SYNTHETIC FUEL is made from a variety of chemical substances. Some synthetic fuels are coal, natural gas, biomass, OIL SHALE (rock containing oil), and bituminous sands that contain material from which oil is recovered. For example, several South African factories make gas from coal.

In Alberta, Canada, an abundance of bituminous sand is used to produce oil. In Brazil, biomass from cassava plants and sugar cane pulp fuels cars. In the United States, some cars run on GASOHOL—a blend of gas and alcohol made from corn or wheat. [*See also* ACID RAIN; AIR POLLUTION CONTROL ACT; ALASKA PIPELINE; ALTERNATIVE ENERGY SOURCES; BONNEVILLE POWER ADMINISTRATION; BREEDER REACTOR; CARBON MONOXIDE; CATALYTIC CONVERTER; CHERNOBYL; CLEAN AIR ACT; COGENERATION; EXXON VALDEZ; FEDERAL ENERGY REGULATORY COMMISSION (FERC); FUEL WOOD; GLOBAL WARMING; MINING; RECLAMATION ACT OF 1902; SMOG; and SYNTHETIC FUEL.]

Fuel Wood

▶A plentiful combustible material used to provide heat. When partially burned with an amount of OXYGEN, fuel wood forms charcoal.

In the past, wood supplied the power for steamboats and trains and for the steam engines used in MINING and manufacturing. As late as the 1940s, many people in the United States still used fuel wood to heat their homes. During the 1970s, many people in industrialized areas purchased wood stoves for use in place of oil or electric heat sources. Although this fuel source conserves FOSSIL FUELS, it creates environmental problems in the form of AIR POLLUTION. Smoke from burning wood hangs in the air, creating smoglike conditions. Because of the smoky conditions created by burning, only a small percentage of people in the United States heat their homes with wood. In many developing countries, fuel wood is the main source of fire for heating and cooking. In some areas, such as sub-Saharan Africa, shortages of fuel wood force people to gather all available wood, contributing to the problems of DEFORESTATION and DESERTIFICATION. [*See also* ALTERNATIVE ENERGY SOURCES; DEFORESTATION; and FOREST.]

◆ At one time, wood warmed many homes in the United States and provided energy for cooking. Today, most homes are heated by electricity, oil, or natural gas.

Fungi

▌The **kingdom** of organisms that includes mushrooms, molds, mildews, **yeasts**, rusts, smuts, and LICHENS. Fungi are everywhere—in water, in air, on damp basement walls, in gardens, and even on foods. A few SPECIES of fungi are parasites that live on or in other organisms. Some fungi, such as mushrooms, may be large, bright, and colorful. But there are many other fungal species, such as yeasts and some molds, that cannot be seen with the unaided eye.

Most species of fungi play a key role in ECOSYSTEMS. Together with BACTERIA and other microscopic organisms, fungi break down, or decompose, organic matter in the ENVIRONMENT. They recycle nutrients from the matter back into the SOIL for use by PLANTS. These DECOMPOSITION activities are as necessary to the continued existence of ecosystems as are the PHOTOSYNTHESIS activities of plants.

Early **classification** schemes grouped fungi with plants. Fungi were grouped this way because, like plants, many fungi do not move about freely, but instead grow anchored in soil or on decaying logs. However, unlike plants, fungi lack chlorophyll and cannot produce their own food. Fungi must therefore obtain the nutrients they need from the environment. As biologists learned more about fungi, they realized that fungi should be classified in a distinct kingdom.

◆ Hyphae, the basic structures of fungi, grow to form extensive fibrous networks called *mycelia.*

STRUCTURE AND FUNCTION OF FUNGI

A few species of fungi, such as the yeasts, are made up of only a single cell. However, most species of fungi are multicellular organisms. The main structures of multicellular fungi are the *hyphae* (singular, *hypha*). Hyphae are thin, threadlike structures that are barely visible to the unaided eye.

As a fungus grows, hyphae branch out to form an extensive mass called a *mycelium* (plural, *mycelia*). With a hand magnifying lens, it is possible to see the individual hyphal threads in the molds that grow on breads. However, the hyphae that make up mushrooms are much more difficult to see

because they are tightly packed into a dense mass.

In many species of fungi, hyphae are divided by complete or partial cell walls. The cell walls separate hyphae into individual cells. Unlike plants, which have cell walls made of cellulose, the cell walls of most fungi contain chitin. Chitin is a a chemical substance found in the external skeletons of lobsters, crabs, INSECTS, and spiders.

HOW FUNGI OBTAIN FOOD

You have probably seen molds growing on bread or fruit. Fungi use these foods for nutrients, just as people do. However, the way fungi digest and take in their food differs from the way you and other animals carry out these processes. Fungi obtain nutrients through extracellular digestion. In this process, food is digested outside of the cells or "body" of the fungus. For example, in a mold-covered slice of bread, hyphae grow over and into the bread. The fungus releases a digestive chemical through its hyphae that digests, or breaks down, the bread into smaller molecules. Once digested, the tiny nutrient molecules are absorbed through the hyphae. The molecules then enter the mold's cells a little at a time.

The digestive process of fungi causes spoilage of foods. However, fungi perform a necessary role in ecosystems. Fungi, along with bacteria and many species of protists, are DECOMPOSERS, organisms that break down complex organic substances into simpler substances that can be used by other organisms. As decomposers, fungi help to clear dead plants, animals, and other organic wastes from the environment. In the process, they return important nutrients, such as nitrogen and phosphorus, to the SOIL.

HOW FUNGI REPRODUCE

Reproduction is the process by which individual organisms continue their species. Fungi have many ways of reproducing. Most species of fungi carry out both sexual and asexual reproduction. Often fungi reproduce sexually to form spores—tiny, reproductive cells that can grow into a new organism. When a spore lands in a place that has all the conditions needed for survival, a threadlike hypha emerges from it. The hypha then branches to form a mycelium. Eventually, some hyphae grow upward and form a spore-containing structure called a *sporangium* (plural, *sporangia*). Mushroom caps are one type of sporangium. The tiny, black dots you might see on bread mold are another type of sporangium.

Fungi have evolved a variety of strategies for dispersing their spores into new areas. Because spores are small and light in mass, they are easily carried by the wind and by other organisms. Some fungal species, such as the puffballs, release millions of spores that can be blown long distances by the slightest breeze. Fungal spores can also be dispersed by water and by animals, such as BIRDS and insects. An example of spores that are carried by animals are those of the fungus that causes Dutch elm disease. The spores of this fungus stick to the bodies of bark beetles that feed on Dutch elm trees. When the beetle moves from an infected tree to a healthy tree, it transfers spores in the process.

CLASSIFICATION

Fungi are a very abundant and diverse group of organisms. Mycologists, scientists who study fungi, have described and identified about 50,000 species so far. Many of these scientists believe as many as 100,000 to 200,000 species of fungi exist. Those that are known are

◆ Bracket fungi are formed on tree trunks.

divided into four distinct groups, called *divisions*. These groups are classified according to how the fungi produce spores and the structures used in this process.

Zygomycetes

If you have ever thrown away bread covered with black dots and a thick fuzz, you are already familiar with one member of the division *Zygomycota*—the common bread mold. Fungi in this division can reproduce through both asexual and sexual means. The group is named for their spores, which are called *zygospores*. Most species of zygomycetes are decomposers.

Ascomycetes

The ascomycetes, or sac fungi, are the most numerous group of fungi. Ascomycetes include the single-celled yeasts, the many-celled mildews, and the fungus that causes Dutch elm disease. The sac fungi get their name from their sac-like structure called *asci*, in which they produce their spores. All sac fungi produce spores in tiny sacs. When they are filled with spores, the sacs break open, releasing their cargo. The spores are then dispersed by wind, water, or animals.

Basidiomycetes

The basidiomycetes, or club fungi, are probably the fungi most familiar to people. Mushrooms, puffballs, and bracket fungi are all basidiomycetes. Other, less common fungi included in this group are the rusts and smuts. The basidiomycetes get their name from their club-shaped reproductive structures, which are called *basidia* (singular, *basidium*). The clublike structures that are located under the caps of mushrooms and toadstools or under the "shelves" of bracket fungi contain spores.

Some species of basidiomycetes are involved in a special partnership with species of ALGAE. The algae and fungi live in such close relationships they are considered a single organism called a *lichen*. The photosynthetic algae receive protection from the fungi by living within its tissues. In return, the algae provide fungi with the energy they need for survival. This is accomplished through the photosynthesis carried out by the algae.

Deuteromycetes

The deuteromycetes are one of the largest divisions of fungi. This division is sometimes called the imperfect fungi. It includes fungi for which a sexual reproduction process has not been discovered.

The deuteromycetes are of economic importance. In particular, fungi in this group are used in the food industry, to flavor cheeses, candies, and soft drinks.

◆ Fungi are used to make blue cheese.

THE IMPORTANCE OF FUNGI

Many species of fungi are edible and are a very important food source in many parts of the world. Some mushrooms, such as the shiitake, and truffles, a species of ascomycete, are considered delicacies in some societies.

Other species of fungi are involved in the food-making process. Perhaps the best known are the yeasts, which are used for brewing alcoholic beverages and for baking. Yeasts used in baking produce the CARBON DIOXIDE that causes bread dough to rise and take on a light, airy texture.

Fungi are also valuable in medicine. Many important medicines, such as the antibiotics penicillin and cyclosporine, have been developed using fungi. Today, yeasts are important tools in the field of genetic research. Research with yeasts has led to the development of a vaccine for hepatitis B. It may one day lead to innovative treatments for such diseases as CANCER and AIDS.

Fungi are also important to the medical field because of the diseases they cause. For example, thrush is a fungal disease of the mouth often seen in very young children and people with AIDS. Athlete's foot and ringworm are skin diseases caused by fungi.

Fungi also cause diseases in plants and are therefore important to the agricultural industry. In the 1840s, a water mold devastated the potato crop in Ireland. The damage was so severe, it caused a FAMINE. Chestnut blight and Dutch elm disease are also caused by fungi. To protect their crops from such diseases, many farmers spray FUNGICIDES on their crops to prevent damage caused by infestations. [*See also* CHEMICAL CYCLES; DECOMPOSITION; FUNGICIDE; and PARASITISM.]

Fungicide

◗ A chemical used to eliminate FUNGI. Fungi are single-celled and many-celled organisms that have cells with walls made of chitin. Examples of fungi are mushrooms, rusts, and molds. Many-celled fungi have fibrous structures. These organisms cannot make their own food. Instead, they feed on other organisms, either dead or alive. If the organisms a fungus feeds upon are alive, the fungus is a *parasite*. If the organisms a fungus feeds on are dead, the fungus is a *saprophyte*. Some fungi attack crops, causing plant disease or food spoilage. Some also attack animals, as parasites, and can cause disease.

FUNGI THAT CAUSE PLANT DISEASES

Most parasitic plant fungi live on or in the PLANT during part of their life cycle and in the SOIL during the rest of it. The survival of fungi depends on the temperature; most require warm and moist conditions. It must also have the right host plant and be in the right part of the plant, such as the root, stem, leaf, flower, seed, or bulb. For example, the powdery mildew that attacks roses infects only the buds and leaves, not the roots. The apple scab fungus infects only the leaves and fruits.

FUNGICIDES USED TO FIGHT PLANT DISEASES

Fungicides used to protect plants from disease may enter the plant's leaves or roots and travel through the plant's tissues. They may also be applied to and work on only selected areas on a plant.

Almost all fungicides disrupt the invading fungus's chemical processes. For example, sulfur compounds containing CARBON, such as thiram or zineb, interfere with proteins produced by the fungus. Other fungicides include metalaxyl and benomyl, and COPPER compounds, such as Bordeaux mixture.

Fungicides are applied to plants in different ways. Some are placed into the soil for the root to absorb. They may be placed on the plant leaf, stem, or flower as sprays or dusts. Fungicides may also be applied to seeds and bulbs.

ALTERNATIVES TO FUNGICIDE USE

Alternatives to using fungicides include using BIOLOGICAL CONTROLS. The use of biological controls consists of using microorganisms that can kill a fungus or prevent its growth in the soil or on the plant. Often, other fungi are used for this purpose. Other alternatives include collecting and destroying infected plants, rotating crops, and treating plant wounds so that fungi cannot enter the the plant through the wounds.

FUNGICIDES USED TO FIGHT ANIMAL DISEASES

Some fungi cause diseases in animals. Fungal diseases of people include athlete's foot, ringworm, and fungal pneumonia. Prevention of athlete's foot depends on having healthy skin, and keeping skin from being too moist. Prevention of ringworm involves avoiding contaminated objects. Treatment of both athlete's foot and ringworm involves using an antifungal ointment for the skin or a medicine, griseofulvin, taken orally. Fungal pneumonia does not have any effective control measures.

PROBLEMS ASSOCIATED WITH FUNGICIDES

Improperly applied and used, fungicides may be harmful to humans, pets, and WILDLIFE. Fungicide labels must be carefully studied, and their warnings must be followed. Some of the precautions include keeping fungicides away from food and drinking water and avoiding skin contact. One concern to keep in mind about fungicides and wildlife is that fungicides used to control plant disease can be picked up by animals and harm them. [*See also* AGRICULTURAL POLLUTION; HEALTH AND NUTRITION; and PARASITISM.]